AF614694

METHODS IN MOLECULAR BIOLOGY™

Series Editor
John M. Walker
School of Life Sciences
University of Hertfordshire
Hatfield, Hertfordshire, AL10 9AB, UK

For further volumes:
http://www.springer.com/series/7651

NanoBiotechnology Protocols

Second Edition

Edited by

Sandra J. Rosenthal and David W. Wright

Department of Chemistry, Vanderbilt University, Nashville, TN, USA

Editors
Sandra J. Rosenthal
Department of Chemistry
Vanderbilt University
Nashville, TN, USA

David W. Wright
Department of Chemistry
Vanderbilt University
Nashville, TN, USA

ISSN 1064-3745 ISSN 1940-6029 (electronic)
ISBN 978-1-62703-467-8 ISBN 978-1-62703-468-5 (eBook)
DOI 10.1007/978-1-62703-468-5
Springer New York Heidelberg Dordrecht London

Library of Congress Control Number: 2013938838

Printed on acid-free paper

Humana Press is a brand of Springer
Springer is part of Springer Science+Business Media (www.springer.com)

Preface

Nanobiotechnology holds the promise of providing revolutionary insight into aspects of biology ranging from fundamental questions such as elucidating molecular mechanisms of brain disorders to extraordinary applications such as the detection of a single cancer cell in a population of a million cells. Since the publication of the first volume of Nanobiotechnology Protocols 6 years ago, this relatively new field at the intersection of nanoscience and biotechnology has advanced from childhood to adolescence. It is already clear, though, that nanobiotechnologies have found a permanent place in the laboratory. To that end, it is essential that the underlying approaches be based on solid, reproducible methods. It is the goal of *NanoBiotechnology Protocols II* to provide novice and experienced researchers alike a cross section of the methods employed in two foundational areas of nanobiotechnology: imaging and detection. *NBPII* also explores new nano-bio constructs and examines the toxicology of nanomaterials.

Nowhere has nanobiotechnology made a more significant impact than in the area of biological imaging and detection. Chapters 1–9 address different nanoprobes and methods for imaging or detection. The substances that make up these probes range from semiconductors to noble metals to carbon and can offer signals several-fold brighter and substantially more stable than traditional organic probes. Additionally, these imaging approaches can be translated to diagnostic approaches for the detection of important pathogens or disease states. Nanobiotechnology is also embracing the interface between hard inorganic materials and the soft functional biology through important designs using amino acids, peptides, and nucleotides (Chapters 10–13). These building blocks afford the research a variety of approaches that not only can be tuned to control the shape, habit, and properties of a nanoparticle but also allow the self assembly of these particles into larger, functional constructs. Finally, in any emerging technology, there are important safety concerns about the materials. This is especially true in an emerging field such as nanobiotechnology, where truly new materials will be interacting in a biological context. For that reason, establishing the toxicology of these materials is an important "next step" in their development and two protocols for toxicity studies are presented in Chapters 14 and 15. Together, these chapters highlight important current areas of development and directions of the field for the future.

Nashvillie, TN, USA *Sandra J. Rosenthal*
David W. Wright

Contents

Contributors

NICHOLAS M. ADAMS • *Department of Chemistry, Vanderbilt University, Nashville, TN, USA*
JOSHUA M. BARNETT • *Vanderbilt Eye Institute, Vanderbilt University, Nashville, TN, USA*
SARAH C. BAXTER • *University of South Carolina, Columbia, SC, USA*
CARLY JO CARTER • *University of Colorado, Boulder, CO, USA*
JERRY C. CHANG • *Department of Chemistry, Vanderbilt University, Nashville, TN, USA*
DAVIN J. CHERNAK • *University of Illinois at Urbana-Champaign, Champaign, IL, USA*
DAI H. CHUNG • *Departments of Pediatric Surgery and Cancer Biology, Vanderbilt University Medical Center, Vanderbilt University, Nashville, TN, USA*
DAVID E. CLIFFEL • *Vanderbilt University, Nashville, TN, USA*
MATTHEW J. CROW • *Duke University, Durham, NC, USA*
DANIEL L. FELDHEIM • *University of Colorado, Boulder, CO, USA*
EDIE C. GOLDSMITH • *University of South Carolina, Columbia, SC, USA*
HONGLIAN GUO • *Department of Electrical Engineering & Computer Science, Vanderbilt University, Nashville, TN, USA*
AARON HANSON • *University of North Dakota, Grand Forks, ND, USA*
REESE HARRY • *Department of Chemistry, Vanderbilt University, Nashville, TN, USA*
FREDERICK R. HASELTON • *Department of Chemistry, Vanderbilt University, Nashville, TN, USA*
OWEN HENDLEY • *Department of Chemistry, Vanderbilt University, Nashville, TN, USA*
TU HONG • *Department of Electrical Engineering & Computer Science, Vanderbilt University, Nashville, TN, USA*
BRIAN J. HUFFMAN • *University of North Alabama, Florence, AL, USA*
JOSEPH A. IZATT • *Duke University, Durham, NC, USA*
ASHWATH JAYAGOPAL • *Vanderbilt Eye Institute, Vanderbilt University, Nashville, TN, USA*
YUHUI JIN • *University of North Dakota, Grand Forks, ND, USA*
CHRISTINE D. KEATING • *Pennsylvania State University, University Park, PA, USA*
JACQUELINE D. KEIGHRON • *Pennsylvania State University, University Park, PA, USA*
MARC R. KNECHT • *University of Kentucky, Lexington, KY, USA*
JOSHUA L. LIU • *Georgia Institute of Technology, Atlanta, GA, USA*
CATHERINE J. MURPHY • *University of Illinois at Urbana-Champaign, Champaign, IL, USA*
ALINA OWCZAREK • *University of Colorado, Boulder, CO, USA*
CHRISTINE K. PAYNE • *Georgia Institute of Technology, Atlanta, GA, USA*
JOHN S. PENN • *Vanderbilt Eye Institute, Vanderbilt University, Nashville, TN, USA*
JONAS W. PEREZ • *Department of Chemistry, Vanderbilt University, Nashville, TN, USA*
JINGBO QIAO • *Department of Pediatric Surgery, Vanderbilt University Medical center, Nashville, TN, USA*

SANDRA J. ROSENTHAL • *Department of Chemistry, Vanderbilt University, Nashville, TN, USA*
MANISH SETHI • *University of Kentucky, Lexington, KY, USA*
CARRIE A. SIMPSON • *University of Colorado, Boulder, CO, USA*
PATRICK N. SISCO • *University of Illinois at Urbana-Champaign, Champaign, IL, USA*
MELISSA C. SKALA • *Vanderbilt University, Nashville, TN, USA*
JOHN W. STONE • *Department of Chemistry, Vanderbilt University, Nashville, TN, USA*
CRAIG J. SZYMANSKI • *Georgia Institute of Technology, Atlanta, GA, USA*
JOSHUA R. TRANTUM • *Vanderbilt University, Nashville, TN, USA*
ADAM WAX • *Duke University, Durham, NC, USA*
DAVID W. WRIGHT • *Department of Chemistry, Vanderbilt University, Nashville, TN, USA*
ELIZABETH R. WRIGHT • *Emory University, Atlanta, GA, USA*
MIN WU • *University of North Dakota, Grand Forks, ND, USA*
YA-QIONG XU • *Departments of Electrical Engineering & Computer Science and Physics & Astronomy, Vanderbilt University, Nashville, TN, USA*
HONG YI • *Emory University, Atlanta, GA, USA*
QI ZHANG • *Pharmacology Department, Vanderbilt University, Nashville, TN, USA*
JULIA XIAOJUN ZHAO • *University of North Dakota, Grand Forks, ND, USA*
YANG ZHAO • *University of North Dakota, Grand Forks, ND, USA*
GRANT R. ZIMMERMAN • *Department of Chemistry, Vanderbilt University, Nashville, TN, USA*

Chapter 1

High-Aspect-Ratio Gold Nanorods: Their Synthesis and Application to Image Cell-Induced Strain Fields in Collagen Films

Davin J. Chernak, Patrick N. Sisco, Edie C. Goldsmith, Sarah C. Baxter, and Catherine J. Murphy

Abstract

Gold nanoparticles are receiving considerable attention due to their novel properties and the potential variety of their uses. Long gold nanorods with dimensions of approximately 20 × 400 nm exhibit strong light scattering and can be easily observed under dark-field microscopy. Here we describe the use of this light-scattering property to track micrometer scale strains in collagen gels and thick films, which result from cell traction forces applied by neonatal heart fibroblasts. The use of such collagen constructs to model cell behavior in the extracellular matrix is common, and describing local mechanical environments on such a small scale is necessary to understand the complex factors associated with the remodeling of the collagen network. Unlike other particles used for tracking purposes, gold nanorods do not photobleach, allowing their optical signal to be tracked for longer periods of time, and they can be easily synthesized and coated with various charged or neutral shells, potentially reducing the effect of their presence on the cell system or allowing selective placement. Techniques described here are ultimately applicable for investigations with a wide variety of cells and cell environments.

Key words Gold nanorods, Fibroblasts, Cells, Cell culture, Strain, Collagen, Particle tracking, Nanoparticles

1 Introduction

Gold nanoparticles are heavily discussed in modern nanomaterial literature due to their many unique properties and potential uses, including cancer therapy, drug delivery, Raman signal enhancement, imaging, and chemical sensing [1–6]. Such a variety of uses is ultimately due to the electronic nature of the gold nanoparticles; these particles are metallic, containing conduction band electrons, and thus they exhibit collective oscillations of these electrons called plasmons upon irradiation with the proper wavelengths of light [7, 8]. In the case of gold, this translates into strong absorption and elastic scattering in the visible and near-infrared portions

Sandra J. Rosenthal and David W. Wright (eds.), *NanoBiotechnology Protocols*, Methods in Molecular Biology, vol. 1026, DOI 10.1007/978-1-62703-468-5_1,

of the electromagnetic spectrum [7, 8]. Nanospheres of gold, ~4–50 nm diameter, absorb and scatter at ~520 nm one plasmon band. Gold nanorods show two plasmon bands; one band is at ~520 nm, corresponding to the short axis of ~4–50 nm, and the long axis band is tunable from ~600 nm to ~1,500 nm or longer wavelengths depending on the aspect ratio (length/width ratio) of the nanorods [7, 8]. Thus, the optical properties of these nanomaterials are controlled by nanoparticle shape. In addition, the position of the plasmon band is affected by local dielectric constant, and degree of aggregation of the nanoparticles [7, 8].

The versatility of these nanoparticles for biological applications also stems from the relative ease with which the particles can be coated with various biocompatible shells. An electrostatic layer-by-layer method can coat gold nanorods in aqueous solution with many commercially available polyelectrolytes in about 30 min; other coating methods (e.g., silica) can be used to enable more specific surface characteristics, such as those which are advantageous for in vivo targeting for medical applications [9–11].

The efficient light-scattering properties of gold nanorods make them easy to visualize in complex wet matrices, including those that contain cells. This chapter is focused on using the light-scattering properties of gold nanorods to quantitatively measure the strain fields associated with cells as they interact with their soft environment. In particular, we focus on cardiac fibroblasts. Fibroblasts are connective tissue cells of the extracellular matrix (ECM), and are key players in wound healing and the formation of scars. Their migration, proliferation, and remodeling activities, associated with wound healing, have been extensively studied [12]. During the healing process, fibroblasts use a variety of methods to remodel their collagenous ECM environment [13]. However, persistent fibroblast activity can be detrimental, leading for example to hypertrophic scarring. Beyond aesthetic concerns, excessive scar formation can cause functional impairment. Scar formation after a heart attack, while initially necessary, can, over time, result in heart tissue that cannot sustain normal cardiac function [14]. While advances in procedural and medicinal treatments have helped reduce scar formation, there is still a great need to understand a complete picture of fibroblast activity in the ECM and allow researchers to control the mechanisms of scarring. To fully understand the complex mechanochemical transductive pathways associated with fibroblast activity (during scar formation or other remodeling activity), it will be necessary to have a better understanding of how ECM remodeling by fibroblasts is triggered by changes in the local mechanical environment [15].

Investigating such biomechanical events has been done through a variety of methods [16]. These methods often use collagen matrices or films as proxies for the more complex ECM; a number of

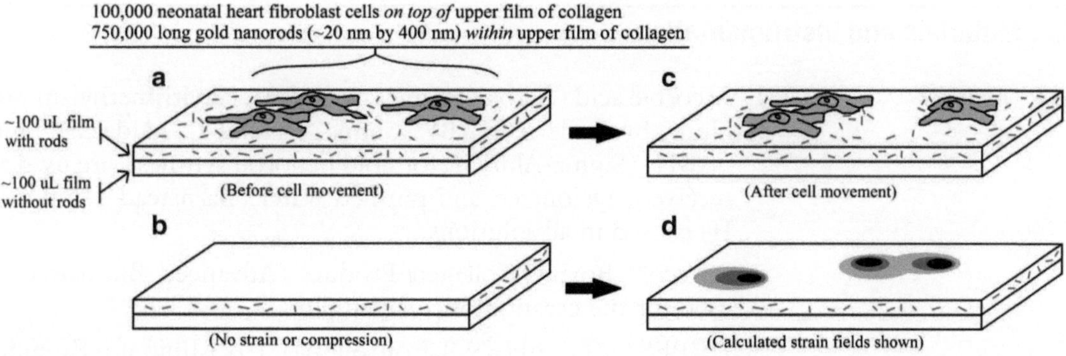

Fig. 1 Fibroblast cells are plated onto a film of collagen embedded with gold nanorods (**a**), and the rods closely associate with the fibers of the collagen network. An image at this initial time serves as a reference state with no strains (**b**). Within hours, the cells attach to the collagen and apply forces, testing and remodeling the local collagen network (**c**). This results in movement of the gold nanorods. Image correlation software tracks the light scattered from the rods, measures the displacement of the nanorods, and calculates the resulting strain fields. The strain fields are illustrated using contour maps (**d**)

techniques involve particle tracking [17, 18]. We have developed a method, using dark-field microscopy, to track the movement of the light scattered from individual gold nanorods in two-dimensional collagen films as the network is remodeled by neonatal cardiac fibroblasts [19]. Using image correlation techniques, this provides us with a measurement of local displacements and approximations of local strain fields within the films. By simultaneously imaging the cells using fluorescent microscopy, these deformations can be matched to cell spatial position and morphology. While cell traction forces can cause strains in collagen gels on the order of 20–36 %, our method has correlated cell movements with collagen strains of less than 0.3 % using image correlation software (VIC-2D©) [20]. An illustration of this process is shown in Fig. 1. The bottom layer of film, without gold nanorods, and the upper layer of film, with gold nanorods, both contain approximately 100 μl of collagen. There are ~750,000 long gold nanorods (20 × 400 nm) embedded *within* the upper film and approximately 100,000 neonatal heart fibroblasts from rats *on top of* the upper film.

The advantage of using gold nanorods in this type of experiment, over other optically active particles, is their relative ease of synthesis and resistance to photobleaching [21]. Herein, we provide details of the technique associated with using gold nanorods as optical trackers of local mechanical fields in cell–tissue constructs, including the synthesis and characterization of long gold nanorods, the isolation and culturing of neonatal heart fibroblast cells, and the preparation and imaging of films of collagen populated with the rods and cells. We describe the interpretation of results using the image correlation software VIC-2D©.

2 Materials and Instrumentation

2.1 Materials

1. Ascorbic acid (Acros Organics or Sigma), cetyltrimethylammonium bromide (CTAB) (Sigma), $HAuCl_4$ (Aldrich), and $NaBH_4$ (Sigma-Aldrich) for gold nanorod synthesis are used as received. Deionized and purified water (Barnstead Nanopure II) is used in all solutions.
2. Purecol® Bovine Collagen Product (Advanced Biomatrix) is used for the creation of collagen films.
3. HEPES buffer (pH = 9.0, from Sigma), 10× Minimum Essential Medium (MEM) (Sigma), culture media*, trypsin solution, * Lab-Tek chamber slides (Thermo Fisher Scientific), and cell culture flasks are required for cell culturing.
4. 70 % by volume ethanol solution.
5. 10 % by volume bleach solution.
6. Serological pipettes.
7. Sterile plastic micropipette tips.
8. Sterile glass aspiration pipette tips.
9. 2 mL cryovials.
10. 15 mL centrifuge tubes.
11. Freezing media*.
12. 5-chloromethylfluoresceindiacetate (CMFDA) or other cell-straining fluorescent dye (Invitrogen).

* *See* Subheading 4 for a description of the components of these mixtures.

2.2 Instrumentation

1. Stir plate.
2. Centrifuge.
3. Class I or class II biosafety cabinet.
4. Transmission electron microscope.
5. Zeta potential phase-analysis light-scattering instrument (e.g., Brookhaven Instruments Zeta PALS).
6. UV–vis spectrophotometer.
7. VIC-2D© Software (Correlated Solutions, http://www.correlatedsolutions.com).

3 Methods

3.1 Synthesis and Characterization of Long Gold Nanorods

Different syntheses of gold nanorods are possible, including those requiring and not requiring different types of seed than what is presented here, those requiring and not requiring silver, those which produce short and long rods, and other photochemical methods [9, 22–25]. Long gold nanorods scatter more light than smaller gold nanorods and so have an advantage when compared to the smaller rods with dark-field imaging. The procedure for the synthesis of the long gold nanorods using gold seeds capped with cetyltrimethylammonium bromide, CTAB, is therefore presented below. A solution of gold seeds is first prepared as described in Subheading 3.1.1, and this solution of seeds is used in the preparation of the long gold nanorods capped in CTAB as described in Subheading 3.1.2. CTAB-capped long gold nanorods greatly decrease cell viability when in the direct presence of cells, and gold nanorods coated with various other polyelectrolytes can potentially induce phenotypic changes in cells [26, 27]. However, the films of collagen used in this procedure maintain rods and cells in different layers, as described in Subheading 3.3, and so use of gold nanorods as is after synthesis, regardless of surface characteristics, has no effect on cell viability.

3.1.1 Preparation of Gold Seeds Capped with Cetyltrimethylammonium Bromide

1. Add 250 μL of 0.01 M $HAuCl_4$ solution to 9.75 mL of 0.1 M cetyltrimethylammonium bromide (CTAB) in a 50 mL centrifuge tube, and stir this solution.
2. In a separate 15 mL centrifuge tube, add 0.0378 × *g* of sodium borohydride to 10 mL of ice-cold, deionized water (18 MΩ is preferable) as quickly as possible. The concentration is now 1.0 M $NaBH_4$.
3. Transfer 1.0 mL of the 1.0 M sodium borohydride solution into 9.0 mL of ice-cold, deionized water (18 MΩ is preferable) in a separate 15 mL centrifuge tube as quickly possible. The concentration is now 0.1 M $NaBH_4$.
4. Transfer 600 μL of the 0.1 M sodium borohydride solution to the 50 mL centrifuge tube containing $HAuCl_4$ and CTAB as quickly as possible. Stir this solution for several minutes, and then stop.
5. Leave this solution undisturbed and unstirred for at least 1 h as the gold seeds are formed. This seed solution must be used for preparation of long gold nanorods within several hours. Typical seeds are approximately 4 nm in diameter.

3.1.2 Synthesis of Long Gold Nanorods from Seeds Capped with Cetyltrimethylammonium Bromide

1. Add 9.0 mL of 0.1 M CTAB to two separate 15 mL centrifuge tubes and 90 mL of 0.1 M CTAB to a 125 mL Erlenmeyer flask.
2. Add 250 μL of 0.1 M $HAuCl_4$ solution to each 15 mL tube and add 2.5 mL of 0.1 M $HAuCl_4$ solution to the Erlenmeyer flask, and mix each solution thoroughly.
3. Add 50 μL of 0.1 M ascorbic acid solution to each 15 mL tube and add 0.5 mL of 0.1 M ascorbic acid solution to the Erlenmeyer flask, and mix each solution thoroughly until completely colorless.

4. Add 1.0 mL of seed solution as synthesized in Subheading 3.1.1 to one of the 15 mL centrifuge tubes, and mix thoroughly.
5. Fifteen seconds after adding seed solution to the first 15 mL centrifuge tube, immediately transfer 1.0 mL of solution from the first centrifuge tube into the second centrifuge tube, and mix the solution in the second centrifuge tube thoroughly.
6. Thirty seconds after adding solution from the first centrifuge tube into the second centrifuge tube, immediately transfer the entire contents of the second centrifuge tube into the Erlenmeyer flask, and mix thoroughly.
7. Cover the Erlenmeyer flask with an upside-down beaker as a lid, and let it sit undisturbed overnight to allow the long gold nanorods to settle to the bottom of the flask.
8. The following day, warm the flask from the outside in a warm water bath to dissolve crystallized CTAB. Ensure that the flask is not heavily disturbed to prevent rods from being resuspended into the supernatant. Decant the supernatant by pouring out the entire contents of the flask. The thin film remaining on the bottom of the flask will be composed almost entirely of the long gold nanorods.
9. Forcefully wash rods from the bottom of the flask with 10–20 mL of deionized water. The presence of the rods will be indicated as the water becomes a pale brown color, and this 10–20 mL we call the 1× solution of gold nanorods. Typical syntheses yield nanorods with diameters of approximately 20 nm and lengths of approximately 400 nm.

3.1.3 Characterization of Long Gold Nanorods After Synthesis

Three common ways to characterize gold nanorods after synthesis are transmission electron microscopy (TEM), ultraviolet–visible (UV–vis) spectroscopy, and zeta potential (effective surface charge) analysis. Here we discuss how these techniques are most commonly used to characterize long gold nanorods.

TEM is used to determine the lengths and widths of nanorods with high precision. Long gold nanorods are prepared for TEM imaging by drop-casting solutions containing the gold nanorods onto a copper TEM grid with lacey or holey carbon mesh and allowing the solution to evaporate. Figure 2 shows a typical TEM image of the nanorods, synthesized using the methods in Subheading 3.1.2.

UV–vis spectroscopy can be used to determine whether high-aspect-ratio rods have been successfully synthesized (rather than short rods or other geometric shapes), as well as the concentration of the rods. However, water absorbs at ~1,200 nm, similar to the longitudinal plasmon band of long rods, so the location of the longitudinal band of long rods becomes difficult to determine with this solvent. Fully deuterated water, D_2O, can be used to suspend

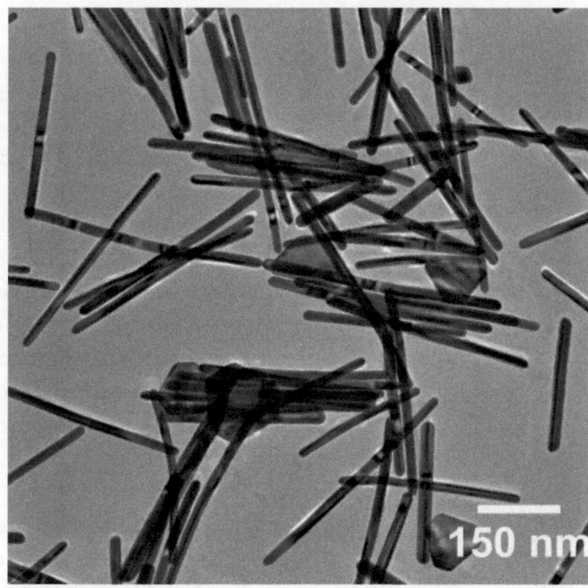

Fig. 2 Typical long gold nanorods as seen under TEM. Dimensions of the rods are near 20 nm by 400 nm

long rods for UV–vis spectroscopy. Centrifuge the rods at 4,000 rcf for 6 min and resuspend them in D_2O. Repeat this centrifugation once more. After the second resuspension in D_2O, quickly take a UV–vis measurement in order to obtain a clean spectrum and avoid the D_2O exchange with humidity in the air that converts it over time to H_2O. Successful synthesis of long nanorods will be indicated by a longitudinal plasmon band near 1,400 nm, although TEM will be necessary to determine the dimensions of the nanorods more precisely.

UV–vis spectroscopy is also used to determine the concentration of gold nanorods by use of the absorbance of the transverse plasmon band at ~520 nm. The usual Beer's law applies:

$$A=\varepsilon bc \qquad A=\varepsilon\times b\times c \tag{1}$$

In Eq. 1, *A* is the absorbance of the transverse plasmon band in the UV–vis spectrum, defined as –log(*T*), where *T*, the transmittance, is the ratio of incident intensity of light over transmitted intensity of light. The path length of solution of gold nanorods, *b*, is 1.0 cm for most UV–vis spectrometer cuvettes but of course depends on the dimensions of the cuvettes used, and the molar extinction coefficient of long gold nanorods, *ε*, is 1.4×10^{10} L mol^{-1} cm^{-1} at ~520 nm (use the peak maximum near that wavelength, in water).

The concentration, *c*, of mol of rods per liter, is then solved by dividing the absorbance of the solution by *b* and *ε*. Typical concentrations of rods after syntheses are on the order of 10^{-10} mol of rods per L.

Zeta potential analysis is used to approximate the surface charge of the synthesized nanorods. The nanorods are coated with a bilayer of CTAB after synthesis and therefore will bear positive charge. Layer-by-layer assembly of polyelectrolytes is a common procedure used with many applications of nanorods (especially for in vivo uses), and measurement of alternating surface charge with each additional layer is used to confirm the application of these layers [9]. However, for our purposes, we will simply use this technique to confirm that the rods are stabilized in solution by a positively charged surface. Zeta potential will vary by solution, but for long gold nanorods suspended in water, typical zeta potential measurements are near +20 mV to +60 mV. Greater magnitudes of zeta potential lead to less aggregation of the rods due to greater repulsions between them.

3.2 Culturing of Neonatal Heart Fibroblast Cells

Neonatal heart fibroblast cells from rats are a biosafety level one hazard, meaning that these nonhuman cells are not known to regularly infect humans and that the cells can be handled in regular laboratory settings with minimal extra precaution. For more details about such safety precautions, we refer readers to Biosafety in Microbiological and Medical Laboratories: Fifth Edition. Sterile environments are preferred however when culturing fibroblast cells in order to avoid contamination of cell cultures and thus acquisition of inadvertently altered data. We therefore recommend that all cell culture procedures, except for centrifugation, thawing of materials in heated baths, freezing of cells in nitrogen, and incubation, be performed in a class II biosafety cabinet. Only presterilized tubes or other containers should be used in this cabinet, pipette tips should be autoclaved before use, and gloves should be sprayed with 70 % by volume ethanol solutions before handling of sensitive materials. Anytime this cabinet is used, it should also be allowed to circulate air for at least 15 min prior to use to sterilize the environment, and work surfaces should be rinsed with 70 % by volume ethanol solutions. Further details about proper use of biosafety cabinets can be found in Primary Containment for Biohazards: Selection, Installation and Use of Biological Safety Cabinets: Third Edition. When culturing cells, decant or remove solutions using a water aspirator with a bleach trap and sterile glass pipette tips. When adding or transferring solutions, use serological pipettes for quantities greater than 1 mL and micropipettes with sterile pipette tips for quantities less than 1 mL.

3.2.1 Thawing of Frozen Fibroblast Cells and Promotion of Their Growth by Plating and Incubation

We receive frozen neonatal heart fibroblast cells from colleagues at the University of South Carolina, and these cells are thawed before use. It is also necessary to thaw cells when removing them from long-term storage in liquid nitrogen (this cryopreservation process is described in Subheading 3.2.4).

Thaw cells quickly by immersing the vial in which they are stored in a heated 37 °C bath until most of the solution is liquid. Our cells are suspended in 1 mL of freezing media containing 10 % dimethyl sulfoxide (DMSO), and so it is preferable to remove them from this solution quickly after thawing to minimize cell death. Transfer the solution of cells into a 15 mL tube prefilled with 12 mL of culture media. The exact volume of culture media is not especially important however; the volume of culture media used is typically around 10 to 15 times greater than the volume of freezing media that is being thawed. Close the tube, and centrifuge it for 8 min at 800 rpm. After centrifugation, remove supernatant, leaving a pellet of fibroblast cells in the bottom of the centrifuge tube. Be careful not to press into the pellet when adding or removing media, as this may kill a large portion of the cells. Add a few mL of media into the tube, and resuspend the pellet entirely by successively redispersing the media with moderate force. Avoid producing bubbles within the solution during this (and any other) process involving cells. After the pellet is dissolved, fill the tube with media to approximately 12 mL.

Open a cell culture flask, and transfer the entire solution from the 15 mL tube into the flask. As best as possible, only dispense the solution onto flask surfaces where cells will plate to maximize cell yield. This may require tipping the flask on its side when dispensing the solution (the flask rests on its side, rather than standing upright, when cells are plating or incubating). When the solution is spread evenly onto the plating surface, incubate the flask at 37 °C and 5 % CO_2 to allow cells to attach and grow. It will take 2–3 days before cells become confluent and nearly cover the plating surface, and it is at this point when the cells can be passaged as described later to increase cell numbers or used for an experiment.

3.2.2 Cell Culture of Fibroblast Cells Before Use in Collagen Films

Cells are considered 100 % confluent when they appear extended and attached to the entire plating surface and there is no more room for cells to grow. At this point cells are ready for use. If cells are too sparse based on the number of cells necessary for the experiment, replace culture media with new media and allow cells to grow for more time to increase cell yield. This is especially important if cells are using up the media and it is turning slightly orange (media is normally red). Overgrowth of cells may result in difficulty removing the cells from the plating surface and the presence of cell clumps rather than individually dispersed cells.

During cell culture, first remove media by standing the flask upright and removing media with a pipette that is inserted into the flask away from the cells. The plated cells will remain attached to the plating surface after removing media. Trypsin is used to lift the cells off of the culture surface, but the serum in the residual media will deactivate the trypsin. Therefore, the residual media is washed out before trypsinization using 1× PBS. Add enough trypsin to cover the cell culture area (this is approximately 2 mL of trypsin for a 250 mm cell culture flask) while tilting the flask and dispensing directly onto the plating surface. Gently swirl the flask to ensure that trypsin entirely covers the plating surface, close the flask, and incubate it at room temperature for 5 min.

After 5 min of incubation of the flask with trypsin, look at the cells under optical microscope. Cells will be balled up and floating in the trypsin rather than extended and held rigidly on the plating surface. If some cells still remain attached to the flask, incubate the flask for more time and possibly elevate incubation temperature to 37 °C, but be careful to not leave cells in this state for too long, because they are currently not suspended in nourishing media and are very stressed by the combination of trypsin and increased temperature. Overexposure to trypsin can kill the cells. Gently tap the flask to detach any cells still stubbornly attached on the plating surface into the trypsin. Also ensure that cells do not experience trypsinization 2 days in a row, i.e., passage cells at most every 2 days.

Add approximately 5 mL of media for every 1 mL of trypsin into the flask, and wash it around to ensure that all cells are washed into media. Transfer the entire contents of the flask into a 15 mL centrifuge tube, and centrifuge the tube for 8 min at 800 rpm. After centrifugation, remove supernatant, leaving the pellet of fibroblast cells in the bottom of the centrifuge tube. Be careful not to press into the pellet and kill cells when adding or removing media. Add a few mL of media into the tube, and break up the pellet entirely by moderately forceful redispersion of the media. Do not add any more media to the tube after breaking up the pellet until after counting the cells.

A hemacytometer is used to count the cells. Transfer 10 μL of solution from the 15 mL tube into the hemacytometer. The solution spreads across several regions in the hemacytometer, and the cells in five of these regions are counted. Count the total number of cells in these five regions, divide this number by five, and multiply it by 10^4 to find the number of cells per mL in the 15 mL tube. With the concentration of cells known, the cells are ready for use in films of collagen, each of which will contain approximately 100,000 cells. We refer readers to Subheading 3.3 for a description of the preparation of films of collagen containing gold nanorods and fibroblast cells.

3.2.3 Short-Term Storage and Multiplication of Fibroblast Cells by Replating, Splitting Flasks

After using cells for films of collagen as described in Subheading 3.3, it is likely that a considerable number of cells from a culture will remain unused. Large confluent cell culture flasks contain millions of fibroblast cells, while each film of collagen uses only 100,000–200,000 cells. Unused cells can be replated for future use as follows.

Add media into the tube that contains cells until about 12 mL of solution are in the tube. Invert the tube a few times to evenly disperse the cells, and transfer the entire solution of the tube into a cell culture flask. As best as possible, only dispense onto surfaces onto which cells will be plating to maximize cells yield. When the solution is spread evenly on the plating surface, place it into incubation at 37 °C and 5 % CO_2 to allow the cells to attach and grow into a large number of new cells.

The number of cells in the flask at this point depends on the number of unused cells remaining after preparation of films, so the length of incubation time before confluence occurs will vary. A greater number of cells become confluent more quickly than a smaller number of cells. Ensure that media which is being used up and turning orange is replaced with fresh media. When cells are confluent, one can trypsinize them as described in Subheading 3.2.2 and then use them for an experiment as in Subheading 3.3 or prepare them for cryopreservation as described in Subheading 3.2.4.

An additional option to increase the number of cells is to split one flask into multiple flasks. In this case, simply trypsinize cells from a flask surface, and suspend them into a few mL of media in a 15 mL centrifuge tube as usual. Now however, transfer half (or a third or fourth) of this solution into another 15 mL centrifuge tube (or two of three tubes), and fill each tube up to 12 mL with media. Now plate the cells from each different tube as usual by dispensing each different solution into a separate flask. By splitting flasks, you can dramatically increase the number of cells in your possession, but be aware that cells should not be used for experiments if they are passage 5 or greater (i.e., if they have experienced trypsin five or more times). Cells with progressively higher passage numbers above five behave progressively more different than lower passage cells. Additionally, only split from one flask into two to five flasks; splitting them to lower concentrations leads to significantly greater incubation time before confluence and often reduced viability.

3.2.4 Long-Term Storage of Fibroblast Cells by Cryopreservation at Liquid Nitrogen Temperatures

Cells that will not be split into new flasks or used within several days can be stored indefinitely in liquid nitrogen. In order to freeze cells, first culture them using the method described above in Subheading 3.2.2; however, after counting the cells split the cells into multiple tubes containing 4–6 million cells. These tubes should be centrifuged at 800 rpm for 8 min. and the resulting pellet is redispersed in 1 mL of freezing media. Transfer this solution of cells and freezing media into a cryovial and store it at −80 °C for at least 24 h. After at least 24 h of storage, place the vial into a liquid nitrogen storage unit. Pertinent information with which to label the cryovial includes the type of cell, its passage number, and the date of storage. Cells can be stored indefinitely at this temperature, and removal of the cells from cryopreservation can be done using the thawing procedure described in Subheading 3.2.1.

3.3 Preparation of Films of Collagen Containing Long Gold Nanorods and Fibroblast Cells

The following procedure is used to produce approximately 1 mL of neutralized collagen, which can be used to create several films, each of which uses between 50 and 250 μL of neutralized collagen. The procedure for making neutralized collagen can easily be scaled up for production of greater numbers of films by using larger quantities of reactants and larger tubes. Produce as much neutralized collagen as needed to create the desired number of films, as well as 1 mL extra to account for the stickiness of the collagen to the sides of tubes in which it is produced.

1. Add Purecol® bovine collagen product to a 1.5 mL centrifuge tube, HEPES buffer (pH = 9) to a second 1.5 mL centrifuge tube, and 10× MEM to a third 1.5 mL centrifuge tube. By placing adequate amounts of each product into these tubes, you can now handle the tubes in order to acquire each product and thus reduce the need to continually handle and possibly contaminate the larger stock products. Replace the Purecol® product to refrigeration immediately when the stock bottle is not in use.
2. Combine 100 μL of ice-cold HEPES buffer with 100 μL of ice-cold MEM in a 1.5 mL centrifuge tube.
3. Layer 800 μL of ice-cold Purecol® product on top of the HEPES/MEM solution and invert rapidly, without introducing air bubbles, to mix the solutions. If prepared properly the solution will turn from a dark pink/fuchsia color to a light rose color. If this color change does not occur, discard and prepare again. The pH of this final collagen solution is near 7.0 and should be kept on ice while preparing films as quickly as possible to avoid collagen polymerization.

Evenly spread 50–250 μL of neutralized collagen into a square film within the chamber of a Lab-Tek chamber slide. Ensure that collagen does not contact the side of the chamber to prevent disturbance of collagen deformation. After the collagen is spread, slightly tilt the slide to collect excess collagen to one corner of the film, and remove this excess collagen with a micropipette. Cure the film in an incubator for a minimum of 2 h at 37 °C and 5 % CO_2 before addition of further layers.

After at least 2 h of incubation of the collagen film, apply another identical layer of neutralized collagen on top of the first layer. Before additional incubation however, add 25 μL of a gold nanorod solution which is synthesized as in Subheading 3.1.2 but concentrated by a factor of three by centrifuging 1× nanorod solution and resuspending it into one third of its original volume. This volume of 3× concentrated solution of gold nanorods typically contains hundreds of thousands of rods. Precise numbers of rods can be used if desired by finding the concentration of rods from Eq. 1 and using the appropriate volume. An easy way to add the gold nanorod solution is to add one drop to each quadrant of the collagen film and one in the center. A small volume of highly concentrated rods is used to minimize the volume of solution added to the collagen [26]. After the nanorods have been added, cure the film in an incubator overnight at 37 °C and 5 % CO_2.

The following day, add 100,000 cells onto the collagen film. The cells will plate onto the collagen film within an hour, so place the chamber slide into incubation as this plating occurs. Time point zero is considered to be when cells are adhered to the film surface, which is typically around 1 h after the addition of cells to the film. Thirty minutes before fluorescent imaging of the cells they must be stained with 5-chloromethylfluoresceindiacetate fluorescent dye (CMFDA, excitation wavelength 450–490 nm). Details regarding imaging of cell activity are described in Subheading 3.4.

3.4 Imaging of Cell-Induced Strain in Films of Collagen by Tracking Movement of Long Gold Nanorods

Before imaging, all media should be removed from the chamber slide. Some residual media will necessarily remain, but this will sustain cells during the imaging process. The walls of the chamber slide can then be removed as well as any adhesive holding them in place. A cover slip should then be placed over the collagen film making sure to avoid trapping bubbles between the coverslip and the film; this will protect the collagen gel and cells as well as microscope components. The slide is then ready for imaging. The time resolution that is desirable during imaging will depend on the cells and actions being imaged; 10–20 min proved to be an appropriate time interval to produce a relatively coherent time sequence of deformation due to the cell traction forces applied by the fibroblasts described here. In order to track movement of cells and the corresponding strain in the collagen, two methods of imaging are used concurrently. Dark-field imaging is used to observe the movement of the random pattern of scattered light produced by the long gold nanorods. Fluorescent imaging is used to image the movement of cells, specifically their spatial position, morphology, and spreading or retracting of their processes. The excitation wavelength of 5-chloromethylfluoresceindiacetate (CMFDA) dye used for the fluorescent imaging is 450–490 nm. By overlaying the dark-field and fluorescent images for each time point, VIC-2D© software can be used as described below to track rod movement and infer evolving strain fields over time, thus associating cell position and morphology with the local strain fields.

3.4.1 Overview of Optical Measurement Techniques

A number of software programs, e.g., the proprietary programs MatLab, PhotoShop, and Mathematica and NIH's ImageJ, provide automated methods of image registration. Image registration is the process of transforming different data sets—in the form of digital images—into the same coordinate system. Image registration is a fundamental tool in many aspects of optical measurement and analysis. Depending on the differences between the pairs of images, the software will try to find an optimal match using a set of mathematical rules. Medical imaging has taken advantage of these techniques to register images taken, for example, at different times, from different perspectives or with different imaging techniques, in order to consolidate information, or build in depth and perspective or capture three dimensionality. If translational or rotational motion (rigid body motion) is described, then image registration can be used to measure displacements of the pattern or features.

If there is deformation, e.g., stretching, as in the work described here, then in addition to positional motion, a more complex analysis of the transformational techniques for alignment may be required to determine differences between the images. It may be possible to use one of these image registration techniques, with additional computational tools to measure spatial displacements, though the authors have not attempted this. Reviews of many of the types, applications, and extensions of registration algorithms exist [28, 29]. For biological applications, similar to the experiments described above, a number of groups have used small numbers of micro-beads to examine how single cells deform their surroundings; the smaller number of particles facilitates calculations of displacements by hand. Engineering strains can then be easily approximated along lines of deformation [30–32].

3.4.2 How VIC-2D© Image Correlation Works

Optical measurements of deformation are by examining pictures (digital images) taken during mechanical loading. If, on a plane of a material (2D), a random pattern (light scattered from gold nanorods) is displayed, then any movement or planar distortion will be evident in a corresponding movement or distortion of the pattern. By tracking how the elements of the pattern move, deformation of the surface can be quantified. Computationally, this tracking, or pattern matching between pairs of images, is based on evaluations of the numeric values of grayscale intensities corresponding to surface pattern. In a grayscale image, pixels exhibit grayscale intensities between 0 (black) and 255 (white); intermediate values produce shades of gray.

The basis of 2-D image correlation is the matching of a digital image of a planar surface before loading, *the reference image* (in our case the undeformed image at time point zero), to an image taken of the surface after loading, *a target image* (in our case a deformed image). The surface must have a visibly distinguishable random pattern (the light scattered from the gold nanorods), where different grayscale levels are associated with individual pixels. The rods attach to the collagen fibers and so the visible pattern of light scattered from the rod moves as the collagen shifts position under the influence of cell movement and migration.

The underlying concept is very simple. If individual pixels in the first image can be tracked to their new location in a second image based on the same coordinate system, then the displacement, the distance the pixel in question has shifted, can be calculated.

In this work, the image correlation software VIC-2D© (Correlated Solutions, http://www.correlatedsolutions.com), developed at the University of South Carolina [33, 34], was used. The particular strength of this software is that it is capable of efficiently tracking pixels and measuring a displacement field over a large and detailed patterned area. Additionally, the software approximates the strain fields, gradients of displacements. Details of this software and a comprehensive discussion of digital image correlation can be found elsewhere [35].

Given an initial guess, VIC-2D© matches points in images by comparing grayscale intensities and doing an efficient computational search of the surrounding area. It determines matching positions by minimizing a correlation function. To illustrate this concept, Fig. 3 illustrates an idealized pattern and example intensities associated with the pattern. Figure 4 shows the same pattern after a rigid body translation; translation is considered the trivial deformation.

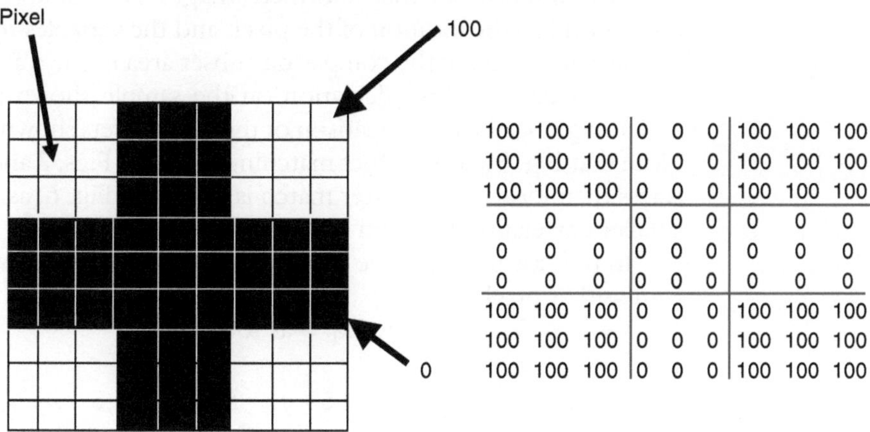

100	100	100	0	0	0	100	100	100
100	100	100	0	0	0	100	100	100
100	100	100	0	0	0	100	100	100
0	0	0	0	0	0	0	0	0
0	0	0	0	0	0	0	0	0
0	0	0	0	0	0	0	0	0
100	100	100	0	0	0	100	100	100
100	100	100	0	0	0	100	100	100
100	100	100	0	0	0	100	100	100

Fig. 3 Idealized undeformed image, with pixel grayscales corresponding to pattern elements registering either 0 (*black*) or 100 (*white*) in intensity (Illustration courtesy of Dr. S.R. McNeill)

100	100	100	100	0	0	0	100	100
100	100	100	100	0	0	0	100	100
0	0	0	0	0	0	0	0	0
0	0	0	0	0	0	0	0	0
0	0	0	0	0	0	0	0	0
100	100	100	100	0	0	0	100	100
100	100	100	100	0	0	0	100	100
100	100	100	100	0	0	0	100	100
100	100	100	100	0	0	0	100	100

Fig. 4 Idealized deformed image, with pixel grayscales corresponding to pattern elements registering either 0 (*black*) or 100 (*white*) in intensity (Illustration courtesy of Dr. S.R. McNeill)

While it is fairly easy to track the deformation between these two images, mapping pixels from Fig. 3 to Fig. 4 by inspection, a method of automating this matching process is obviously required for larger images, more detailed patterns, or more complex deformations. In VIC-2D© this is done by selecting a sequence of small subsets which raster over the original image and calculating a correlation function between each of these areas and all possible similarly sized areas in the deformed image. The region with the minimum correlation value, calculated between each pair of regions, corresponds to the deformed location of the original subset. The correlation function is given in Eq. 2:

$$C(x,y,u,v) = \sum_{i,j=-n/2}^{n/2} (I(x+i,y+j) - I^{*}(x+u+i,y+v+j))^2. \quad (2)$$

In Eq. 2, I is the numeric value of the intensity of the undeformed image at each pixel location, done here at integer locations. I^* is the intensity in the deformed image. The variables x and y correspond to the location of the pixel, and the variables u and v are the displacements of the comparing subset area in the x and y directions. Performing this calculation on the sample shown in Figs. 3 and 4 might lead to a comparison of the two subsets shown in Fig. 5. Note that Fig. 5 shows subset matching between Figs. 3 and 4 which is less than ideal, and a better match is shown in Fig. 6, as is seen by smaller correlation function associated with Fig. 6.

In the case of Fig. 5, the correlation function for the 5 × 5 subset, centered at pixel

$x=5$, $y=5$, is given by Eq. 3 as follows:

$$C(5,5,-2,-2) = \sum_{i,j=-2}^{2} (I(5+i,5+j) - I^{*}(5-2+i,5-2+j))^2. \quad (3)$$

The (−2)'s correspond to the relative position (the center point) of the subset to which the undeformed region is being compared. Expanding this out yields the following:

$$\begin{aligned}&(100-0)^2+(0-0)^2+(0-0)^2+(0-0)^2+(100-0)^2+\\&(0-100)^2+(0-100)^2+(0-100)^2+(0-100)^2+(0-0)^2+\\&(0-100)^2+(0-100)^2+(0-100)^2+(0-100)^2+(0-0)^2+\end{aligned} \quad (4)$$

$$(0-100)^2+(0-100)^2+(0-100)^2+(0-100)^2+(0-0)^2+$$
$$(100-100)^2+(0-100)^2+(0-100)^2+(0-100)^2+(100-0)^2=18{,}000$$

Since it is possible to find the deformed position in this example by inspection, and it is clearly not the one illustrated in Fig. 5, it is not surprising that this subset has a high correlation value. In this idealized example, the correlation function for the real deformed position of this subset is exactly zero, as seen in Fig. 6. Zero is the absolute minimum for the sum of squared numbers.

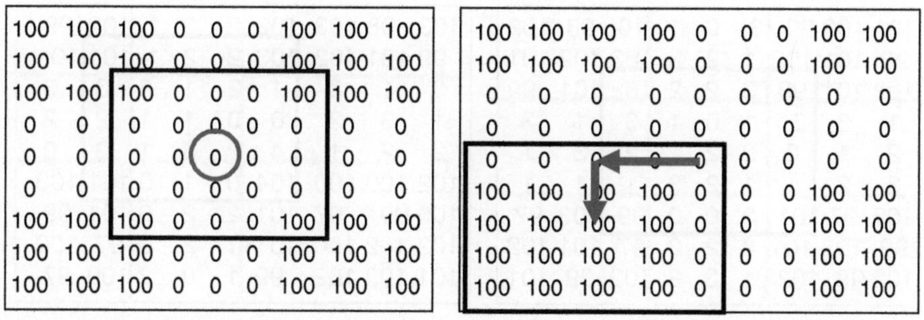

100	100	100	0	0	0	100	100	100
100	100	100	0	0	0	100	100	100
100	100	100	0	0	0	100	100	100
0	0	0	0	0	0	0	0	0
0	0	0	0	0	0	0	0	0
0	0	0	0	0	0	0	0	0
100	100	100	0	0	0	100	100	100
100	100	100	0	0	0	100	100	100
100	100	100	0	0	0	100	100	100

100	100	100	100	0	0	0	100	100
100	100	100	100	0	0	0	100	100
0	0	0	0	0	0	0	0	0
0	0	0	0	0	0	0	0	0
0	0	0	0	0	0	0	0	0
100	100	100	100	0	0	0	100	100
100	100	100	100	0	0	0	100	100
100	100	100	100	0	0	0	100	100
100	100	100	100	0	0	0	100	100

Fig. 5 Image analysis compares correlation function values of subsets in deformed image to find match in undeformed image. The undeformed subset (with *circled center*) is compared to an arbitrary subset in deformed image. Arbitrary deformed area is shifted 2 to the left and 2 down from the circled center point of the undeformed subset. Figure 6 illustrates a better match between these deformed and undeformed images than the one in this figure (Illustration courtesy of Dr. S.R. McNeill)

100	100	100	0	0	0	100	100	100
100	100	100	0	0	0	100	100	100
100	100	100	0	0	0	100	100	100
0	0	0	0	0	0	0	0	0
0	0	0	0	0	0	0	0	0
0	0	0	0	0	0	0	0	0
100	100	100	0	0	0	100	100	100
100	100	100	0	0	0	100	100	100
100	100	100	0	0	0	100	100	100

100	100	100	100	0	0	0	100	100
100	100	100	100	0	0	0	100	100
0	0	0	0	0	0	0	0	0
0	0	0	0	0	0	0	0	0
0	0	0	0	0	0	0	0	0
100	100	100	100	0	0	0	100	100
100	100	100	100	0	0	0	100	100
100	100	100	100	0	0	0	100	100
100	100	100	100	0	0	0	100	100

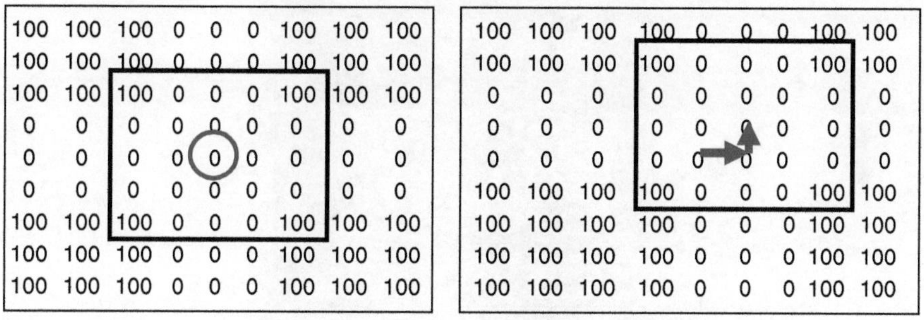

Fig. 6 Image analysis compares a correlation function of subsets in a deformed image to find a match in an undeformed image. Subset with minimum correlation function value, $C(5,5,1,1) = 0$ (Illustration courtesy of Dr. S.R. McNeill)

In real life, the numbers are not as clean and exact as described above. A more realistic example might be more like the one shown in Fig. 7, which includes both image signal and noise.

Although it is still possible to visually track the deformation using the grids shown in Fig. 7, the correlation function value for the match can no longer be exactly zero, as seen in Eq. 5.

$$\begin{aligned} C(5,5,1,1) &= (99-103)^2 + (3-2)^2 + (2-2)^2 + (2-1)^2 + (102-101)^2 + \\ &(2-1)^2 + (1-2)^2 + (0-1)^2 + (1-2)^2 + (3-3)^2 + \\ &(0-0)^2 + (3-0)^2 + (2-1)^2 + (2-1)^2 + (2-1)^2 + \\ &(1-4)^2 + (1-2)^2 + (2-3)^2 + (3-1)^2 + (1-2)^2 + \\ &(101-104)^2 + (3-0)^2 + (0-1)^2 + (0-0)^2 + (99-101)^2 = 71. \end{aligned} \tag{5}$$

Once pixels have been matched, displacements can be calculated. If the coordinate of the first pixel was (1,1) and its position after deformation is (2,3), then its horizontal displacement $u=(2-1)=1$ and its vertical displacement $v=(3-1)=2$. VIC-2D©

101	103	99	3	0	1	100	98	102
99	101	102	1	2	3	103	102	101
102	101	99	3	2	2	102	101	99
1	3	2	1	0	1	3	1	3
2	1	0	3	2	2	2	3	1
3	2	1	1	2	3	1	1	1
102	97	101	3	0	0	99	103	97
99	103	101	2	2	2	97	101	102
102	98	102	1	3	2	102	99	101

103	98	103	102	2	3	1	102	99
99	101	102	103	2	2	1	101	101
3	1	3	1	2	1	2	3	3
1	3	2	0	0	1	1	2	2
2	2	1	4	2	3	1	2	0
102	100	100	104	0	1	0	101	103
100	99	97	101	2	2	0	101	99
102	102	100	101	1	2	3	97	100
101	103	102	99	1	0	2	100	97

Fig. 7 Image analysis compares subsets between undeformed and deformed images to find a match in the undeformed image. Intensities illustrated here include signal and imaging noise (Illustration courtesy of Dr. S.R. McNeill)

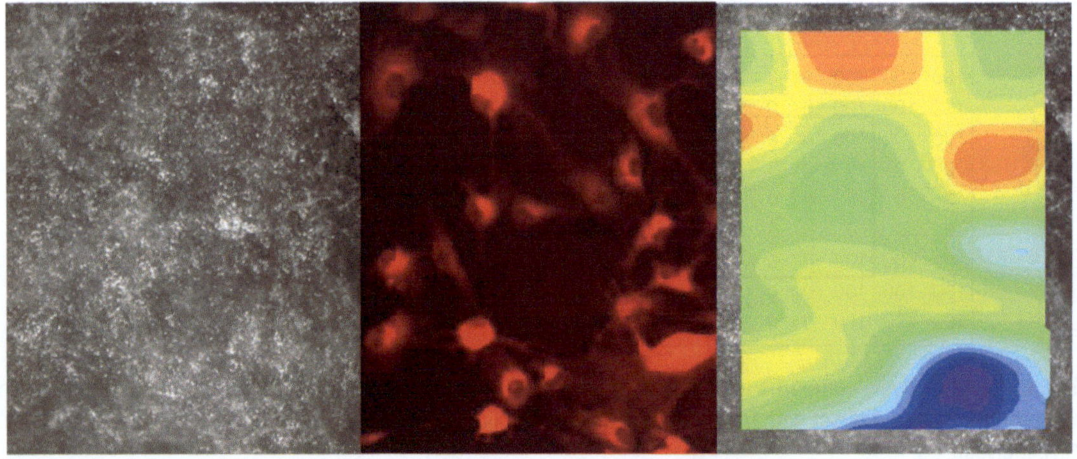

Fig. 8 L–R Grayscale bitmap of pattern of light scatted from gold nanorods, fluorescent image of cells, and vertical strain field map. The field of view is approximately 1 mm

fits an optimal surface through the fields of displacements. Strains, normalizations of displacement, are defined in terms of gradients of displacements which are easily calculated from the surface.

After image acquisition and analysis is complete, information gained from a typical experiment involving collagen remodeling by fibroblast cells is similar to what is seen in Fig. 8 above. The figure shows, from left to right, the grayscale image (deformed) of light scattered from the gold nanorods, a "simultaneous" fluorescent image of cell position and morphology, and the vertical ε_{yy} strains at this time step, as compared to a previous one.

By comparing fluorescent images of cells at different time points, it becomes readily apparent which cell movements are responsible for the correspondingly developed strain fields in the collagen. The collagen strain fields themselves are represented quantitatively in the strain field map, with different colors representing with good precision the amount of strain the collagen experiences. The resolution of this technique is also apparent due to the fact that collagen strains seen here were found to be less than 0.3 % [19]. This fact further demonstrates the sensitivity and applicability of the technique to even minute cell–environment interactions involved with a wide variety of cells and cell environments.

In principle, one could use the optical properties of any submicron particle to track strain fields as described herein. Gold nanorods have the advantage in that their elastic light-scattering properties are not subject to photobleaching and can be broadly excited throughout the visible—in fact, we generally use a simple white light source.

4 Notes

1. Culture media is composed of Dulbecco's Modified Eagle Medium (DMEM), 10 % newborn calf serum (NBCS), 5 % fetal bovine serum (FBS), and penicillin–streptomycin solution (pen. strep).
2. Trypsin solution, referred to in this chapter as simply trypsin, is composed of 0.25 % trypsin and 1 mM ethylenediaminetetraacetic acid (EDTA).
3. Freezing media is composed of 60 % DMEM, 30 % FBS, and 10 % dimethyl sulfoxide (DMSO).

References

1. Huang X, El-Sayed IH, Qian W, El-Sayed MA (2006) Cancer cell imaging and photothermal therapy in the near-infrared region by using gold nanorods. J Am Chem Soc 128(6):2115–2120
2. Paciotti GF, Myer L, Weinreich D, Goia D, Pavel N, McLaughlin RE, Tamarkin L (2004) Colloidal gold: a novel nanoparticle vector for tumor directed drug delivery. Drug Deliv 11(3):169–183
3. Hu X, Cheng W, Wang T, Wang Y, Wang E, Dong S (2005) Fabrication, characterization, and application in SERS of self-assembled polyelectrolyte–gold nanorod multilayered films. J Phys Chem B 109(41):19385–19389
4. Jain PK, Lee KS, El-Sayed IH, El-Sayed MA (2006) Calculated absorption and scattering properties of gold nanoparticles of different size, shape, and composition: applications in biological imaging and biomedicine. J Phys Chem B 110(14):7238–7248
5. Nath N, Chilkoti A (2002) A colorimetric gold nanoparticle sensor to interrogate biomolecular interactions in real time on a surface. Anal Chem 74(3):504–509
6. Murphy CJ, Gole AM, Hunyadi SE, Stone JW, Sisco PN, Alkilany A, Kinard BE, Hankins P (2008) Chemical sensing and imaging with metallic nanorods. Chem Commun 5:544
7. Kelly KL, Coronado E, Zhao LL, Schatz GC (2003) The optical properties of metal nanoparticles: the influence of size, shape, and dielectric environment. J Phys Chem B 107(3):668–677
8. Sosa IO, Noguez C, Barrera RG (2003) Optical properties of metal nanoparticles with arbitrary shapes. J Phys Chem B 107(26):6269–6275
9. Gole A, Murphy CJ (2005) Polyelectrolyte-coated gold nanorods: synthesis. Characterization and Immobilization. Chem Mater 17(6):1325–1330
10. Mandal TK, Fleming MS, Walt DR (2002) Preparation of polymer coated gold nanoparticles by surface-confined living radical polymerization at ambient temperature. Nano Lett 2(1):3–7

11. Durr NJ, Larson T, Smith DK, Korgel BA, Sokolov K, Ben-Yakar A (2007) Two-photon luminescence imaging of cancer cells using molecularly targeted gold nanorods. Nano Lett 7(4):941–945
12. Muir IFK (1998) Control of fibroblast activity in scars: a review. Eur J Plast Surg 21(1):1–7
13. Dallon JC, Ehrlich HP (2008) A review of fibroblast-populated collagen lattices. Wound Repair Regen 16(4):472–479
14. Manabe I, Shindo T, Nagai R (2002) Gene expression in fibroblasts and fibrosis: involvement in cardiac hypertrophy. Circ Res 91(12): 1103–1113
15. Pedersen JA, Swartz MA (2005) Mechanobiology in the third dimension. Ann Biomed Eng 33(11):1469–1490
16. Wang JH, Lin J (2007) Cell traction force and measurement methods. Biomech Model Mechanobiol 6(6):361–371
17. Tseng Y, Kole TP, Wirtz D (2002) Micromechanical mapping of live cells by multiple-particle-tracking microrheology. Biophys J 83(6):3162–3176
18. Levi V (2005) 3-D particle tracking in a Two-photon microscope: application to the study of molecular dynamics in cells. Biophys J 88(4): 2919–2928
19. Stone JW, Sisco PN, Goldsmith EC, Baxter SC, Murphy CJ (2007) Using gold nanorods to probe cell-induced collagen deformation. Nano Lett 7(1):116–119
20. Vanni S, Christoffer Lagerholm B, Otey C, Lansing Taylor D, Lanni F (2003) Internet-based image analysis quantifies contractile behavior of individual fibroblasts inside model tissue. Biophys J 84(4):2715–2727
21. Schultz S, Smith DR, Mock JJ, Schultz DA (2000) Single-target molecule detection with nonbleaching multicolor optical immunolabels. Proc Natl Acad Sci USA 97(3):996–1001
22. Jana NR, Gearheart L, Murphy CJ (2001) Wet chemical synthesis of high aspect ratio cylindrical gold nanorods. J Phys Chem B 105(19): 4065–4067
23. Busbee BD, Obare SO, Murphy CJ (2003) An improved synthesis of high-aspect ratio gold nanorods. Adv Mater 15(5):414–416
24. Liu G-SP (2005) Mechanism of silver(I)-assisted growth of gold nanorods and bipyramids. J Phys Chem B 109(47):22192–22200
25. Kim F, Song JH, Yang P (2002) Photochemical synthesis of gold nanorods. J Am Chem Soc 124(48):14316–14317
26. Zitova B (2003) Image registration methods: a survey. Image and Vision Computing 21(11): 977–1000
27. Wilson CG, Sisco PN, Gadala-Maria FA, Murphy CJ, Goldsmith EC (2009) Polyelectrolyte-coated gold nanorods and their interactions with type I collagen. Biomaterials 30(29):5639–5648
28. Sisco PN, Wilson CG, Mironova E, Baxter SC, Murphy CJ, Goldsmith EC (2008) The effect of gold nanorods on cell-mediated collagen remodeling. Nano Lett 8(10):3409–3412
29. Brown LG (1992) A survey of image registration techniques. ACM Comput Surv 24(4):325–376
30. Wang N, Ostuni E, Whitesides GM, Ingber DE (2002) Micropatterning tractional forces in living cells. Cell Motil Cytoskeleton 52(2):97–106
31. Butler JP, Tolic-Norrelykke IM, Fabry B, Fredberg JJ (2002) Traction fields, moments, and strain energy that cells exert on their surroundings. Am J Physiol Cell Physiol 282(3): C595–C605
32. Tolic-Nørrelykke IM, Wang N (2005) Traction in smooth muscle cells varies with cell spreading. J Biomech 38(7):1405–1412
33. Sutton M, Wolters W, Peters W, Ranson W, McNeill S (1983) Determination of displacements using an improved digital correlation method. Image and Vision Computing 1(3): 133–139
34. Sutton M, Mingqi C, Peters W, Chao Y, McNeill S (1986) Application of an optimized digital correlation method to planar deformation analysis. Image and Vision Computing 4(3):143–150
35. Sutton MA, Ortwu J-J, Schreir, H (2009) Image correlation for shape, motion and deformation measurements; basic concepts, theory and applications. Springer Science + Business Media, LLC.

Chapter 2

Imaging Intracellular Quantum Dots: Fluorescence Microscopy and Transmission Electron Microscopy

Craig J. Szymanski, Hong Yi, Joshua L. Liu, Elizabeth R. Wright, and Christine K. Payne

Abstract

Quantum dots (QDs) and other nanoparticles require delivery and targeting for most intracellular applications. Despite many advances, intracellular delivery and targeting remains inefficient with many QDs remaining bound to the plasma membrane rather than internalized into the cell. The fluorescence resulting from these extracellular QDs results in a background signal that competes with intracellular QDs of interest. We present two methods for the reduction and discrimination of signal resulting from plasma membrane-bound QDs. The first method, a photophysical approach, uses an extracellular quencher to greatly reduce the fluorescence signal from extracellular QDs. This method is compatible with fast, widefield, fluorescence imaging in live cells. Results are presented for two extracellular quenchers, QSY-21 and trypan blue, used in combination with 655 nm emitting QDs. The use of an extracellular quencher can be extended to a wide variety of fluorophores. The second method uses transmission electron microscopy (TEM) to image thin (60–70 nm) slices of resin-embedded cells. The use of sectioned cells and high-resolution TEM makes it possible to discriminate between plasma membrane-bound and intracellular QDs. To overcome the difficulties associated with using TEM to image individual QDs in cells, we have utilized a silver enhancement method that significantly improves the contrast of QDs in TEM images.

Key words Fluorescence microscopy, Transmission electron microscopy, Quantum dot, Quencher, Silver enhancement

1 Introduction

One of the most exciting advances at the interface of nanoscience and biology has been the development of quantum dots (QDs), semiconductor nanoparticles, for applications in cellular imaging [1–6]. QDs are ideal fluorescent probes for cellular imaging due to their brightness, photostability, and ability to be functionalized. Unlike traditional organic or protein fluorophores, QDs offer the possibility of single-molecule imaging, over long timescales, inside living cells. In addition to their use in fluorescence microscopy, the relatively electron-dense core of QDs allows for their use in

Sandra J. Rosenthal and David W. Wright (eds.), *NanoBiotechnology Protocols*, Methods in Molecular Biology, vol. 1026, DOI 10.1007/978-1-62703-468-5_2, © Springer Science+Business Media New York 2013

electron microscopy [7, 8]. Despite these benefits, fluorescent proteins have a considerable advantage for cell imaging as they are highly specific, expressed fused to a protein of interest [9]. In comparison, QDs must be delivered and targeted for most intracellular applications [10–12]. This remains a highly inefficient process with many highly fluorescent QDs remaining bound to the plasma membrane. Overall, this leads to a significant technical concern, how to reduce the fluorescent signal resulting from QDs or other fluorescent nanoparticles that remain bound to the plasma membrane and obscure the intracellular dynamics of interest.

Conventional methods used to exclude the signal from extracellular fluorophores include confocal microscopy and specific washing steps to remove surface-bound fluorophores. Confocal microscopy uses an aperture to remove the out-of-plane fluorescence from an optical slice of a cell [13]. Confocal microscopy requires raster scanning a laser beam or the use of a spinning disc, which limits temporal resolution or z-resolution, respectively. The use of enzymes, such as trypsin, or low pH to remove surface-bound fluorophores has the advantage that these methods can be combined with widefield fluorescence microscopy [14–17]. However these methods are highly dependent on the mechanism by which the fluorophore binds to the cell and are not applicable to all functionalization approaches [18]. Fluorogenic nanoparticles that only emit following binding to a target or within a specific intracellular environment may also remove the extracellular signal, but the design of fluorogenic nanoparticles requires highly specific functionalization for each application [19–24].

While TEM cannot be carried out on live cells, it, unlike fluorescence microscopy, has the resolution to determine the location of individual QDs relative to the plasma membrane. For imaging intracellular QDs, TEM also benefits from standard sample preparation. Cells are routinely processed for TEM by embedding them in a resin and then slicing them into 60–70 nm sections on an ultramicrotome. While the use of gold nanoparticles is well established for cellular TEM, images of QDs in cells have suffered from poor contrast [7, 8]. The semiconductor core of quantum dots has a sufficient electron density for TEM, but this is a relatively low density when competing against the metal salt-stained cellular structures that result following fixation and staining for TEM.

This chapter presents two methods for the improved visualization of intracellular QDs or other suitable fluorescent nanoparticles. Common to both methods is the overall goal of rejecting the signal resulting from QDs that remain on the cell surface. The first method is a photophysical approach using two different extracellular quenchers, QSY-21 and trypan blue (TB), for use with widefield fluorescence microscopy [25, 26]. This enables rapid imaging of QDs without the inclusion of signal from extracellular

QDs. The second method is a physical approach using cell sectioning and silver enhancement for use with TEM. The physical method requires ultramicrotoming of resin-embedded fixed cells to generate thin, 60–70 nm, cell sections. Silver nucleation onto the QDs is used to enhance the QD visibility for high-contrast TEM images.

2 Materials

2.1 Cell Culture and Quantum Dots

1. BS-C-1 cells (ATCC) grown on glass bottom dishes (MatTek Corp., Ashland, MA, USA).
2. Standard cell culture reagents including Minimum Essential Medium (MEM, Invitrogen, Carlsbad, CA, USA) and Fetal Bovine Serum (FBS, Invitrogen, Carlsbad, CA, USA).
3. QTracker 655 Cell Labeling Kit (Invitrogen, Carlsbad, CA, USA).

2.2 Photophysical Method: Extracellular Quencher and Fluorescence Microscopy

1. Quencher. QSY-21 carboxylic acid, succinimidyl ester (Invitrogen, Carlsbad, CA, USA, Cat#: Q-20132) or Trypan Blue (TB, Sigma-Aldrich, St. Louis, MO, USA).
2. Imaging medium of choice.
3. Widefield fluorescence microscope. Images in this chapter were acquired using an Olympus IX-71 with 75 W xenon lamp; Olympus 60× 1.20 NA, water-immersion objective; and an Andor iXon+-enhanced CCD camera. QDs were imaged with a QDLP-A filter set (Semrock). DAPI was imaged with a DAPI-1160A-000 filter set (Semrock).

2.3 Physical Method: Ultramicrotoming and Transmission Electron Microscopy

1. Fixative. 2.5 % glutaraldehyde buffered in 0.1 M sodium phosphate (pH 7.4) and stored at 4 °C.
2. Rinse buffer. 0.1 M sodium phosphate buffer (pH 7.4).
3. Aurion Enhancement Conditioning Solution (ECS, Electron Microscopy Sciences, Hatfield, PA, USA, Cat# 25830).
4. Silver enhancement solution. Aurion R-Gent SE-EM Kit (Electron Microscopy Sciences, Hatfield, PA, USA, Cat# 25521).
5. Enhancement termination solution. 0.03 M sodium thiosulfate in ECS.
6. Post-fixative. 0.5 % OsO_4 in 0.1 M sodium phosphate buffer (pH 7.4).
7. Ethanol series. Ethanol solutions at 25, 50, 70, 95, and 100 % by volume ethanol in water.
8. Epoxy resin. Eponate 12 resin kit (Ted Pella Inc., Redding, CA, USA, Cat# 18010).

9. First counterstain solution. 5 % uranyl acetate in water.
10. Second counterstain solution. 2 % lead citrate in water.
11. Ultramicrotome and glass/diamond knives.
12. Hitachi H-7500 or comparable TEM.

3 Methods

3.1 Binding and Internalization of Quantum Dots

QDs (655 nm emission, Cell Tracker, Invitrogen) were prepared for cellular delivery according to the manufacturer's instructions. For binding to the plasma membrane, cells were cooled to 4 °C and then incubated with QDs at 4 °C for 10 min. The low temperature prevents internalization of the QDs [27]. For intracellular imaging, cells were incubated with QDs for 1 h at 37 °C in a 5 % carbon dioxide environment which resulted in 60 % internalization of QDs (unpublished data).

3.2 Fluorescence Microscopy

QDs were imaged using the fluorescence microscope described above. The use of an extracellular quencher requires no additional optical components. Selection of the proper QD-quencher pair is critical. The selection methods described below can be easily extended to other fluorophores and fluorescent nanoparticles. Issues that must be considered include spectral overlap between fluorophore and quencher, concentration, and cytotoxicity.

3.2.1 Spectral Overlap of Quantum Dot and Quencher

The largest factor in determining which quencher to use is the overlap of the fluorophore emission with the quencher absorption. The emission spectrum of the 655 nm emitting QDs is highly overlapped with the absorption spectrum of QSY-21 and, to a lesser extent, TB indicating that there should be efficient energy transfer from the fluorophore to either of these quenchers (Fig. 1). For QDs or fluorophores that emit at shorter wavelengths, possible quenchers include QSY-7 (Invitrogen) and QSY-14 (Invitrogen).

3.2.2 Optimal Quencher Concentration

The optimal concentration of quencher in the extracellular medium will vary depending on the experiment and the imaging setup. For the experiments described below, 95 % quenching was chosen as the required quenching efficiency. Quenching efficiency is determined as a function of concentration by measuring the fluorescence emission of a constant concentration of fluorophore in the presence of increasing concentrations of quencher (Fig. 2). From this measurement it can be seen that 1 μM QSY-21 is a highly efficient quencher with >99 % quenching of the emission of the 655 nm QDs. TB is a somewhat less effective quencher with 93 % quenching at 1 μM TB.

The upper limit of the quencher concentration is determined by solubility, cytotoxicity, and cost (*see* **Note 1**). TB is water soluble

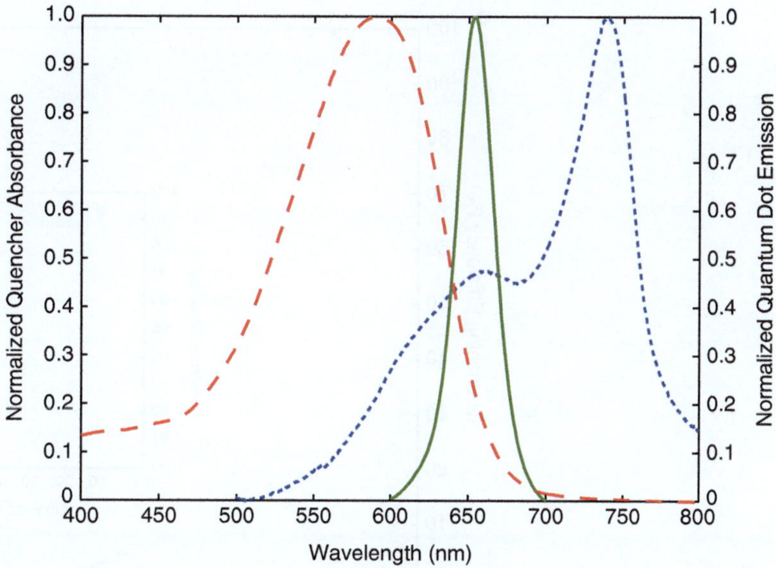

Fig. 1 Normalized emission spectrum of 655 nm QDs (*green, solid*) plotted with the normalized absorption spectra of QSY-21 (*blue, dotted*) and TB (*red, dashed*) in PBS. The greater overlap of the absorption of QSY-21 with QDs leads to more efficient energy transfer than with TB

at a concentration of 100 mM, much higher than those needed for nearly complete quenching of the QDs. QSY-21 is much less soluble in aqueous media and 70–300 nm diameter aggregates can be detected via dynamic light scattering in 1 μM solutions of QSY-21 in PBS. It is not known whether aggregates play a role in the QSY-21 quenching mechanism of QDs, but the aggregates observed in the 1 and 10 μM solutions (300–1,000 nm diameter aggregates) do not generally interfere with imaging. Higher concentrations of QSY-21 in PBS (100 μM) produce large aggregates that are visible by the eye. These aggregates can enter the field of view and interfere with imaging.

Quencher concentration is also limited by the possible fluorescence of the quencher itself. QSY-21 in PBS emits in the region of 670–710 nm (Fig. 3). This emission could be blocked by a 670 nm short-pass filter in the emission pathway. TB is emissive over a much larger range that is highly overlapped with the emission of 655 nm QDs. This signal cannot be filtered out without also filtering the QD signal. This places an upper limit on both the TB concentration and the volume of the quenching medium that can be present above the cells, as it will contribute to out-of-focus fluorescence.

Cytotoxicity can also place an upper limit on quencher concentration. Cell viability studies indicate that neither 10 μM QSY-21 nor TB is cytotoxic for periods of up to 1 h. Additionally, TB is a well-established reagent for the measurement of cell viability.

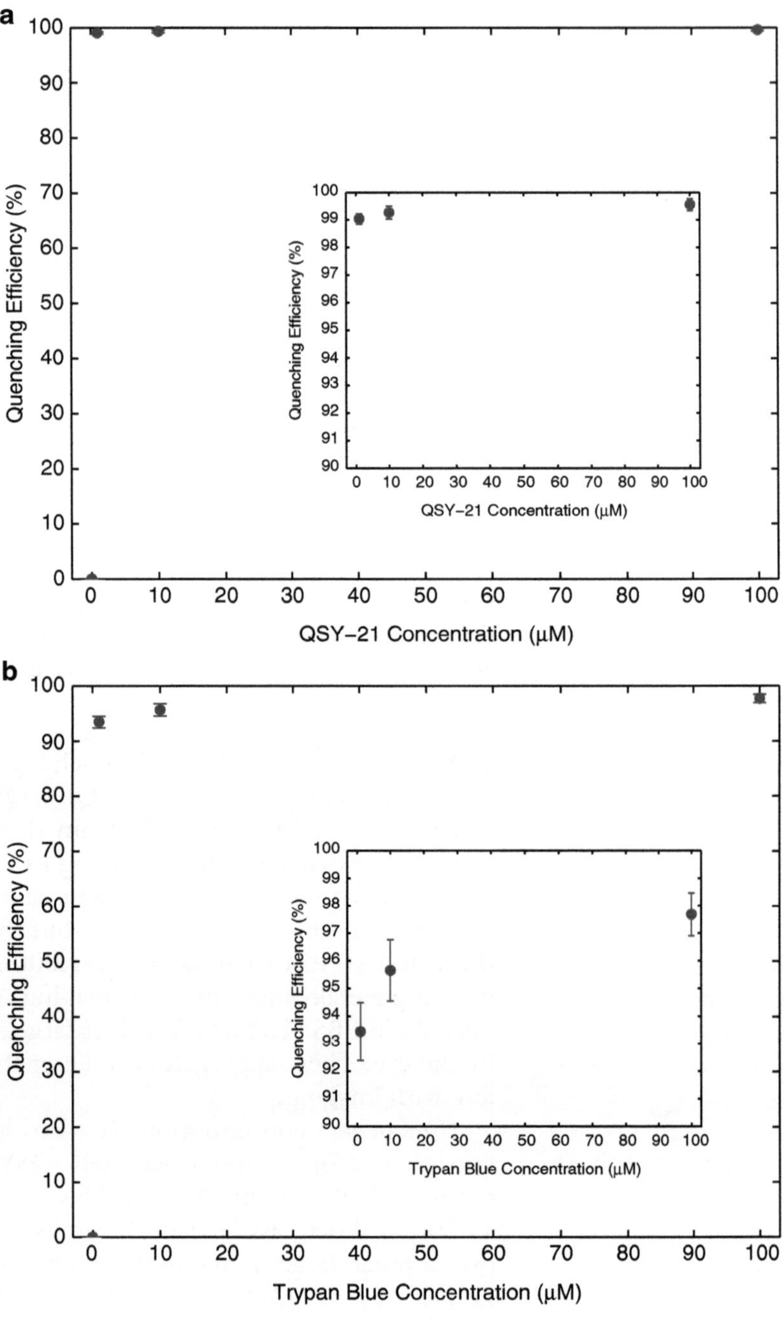

Fig. 2 Effect of quencher concentration on the emission of QDs (4 nM) in PBS. Insets show an expanded view of 1–100 μM quencher. QDs were excited at 450 nm. (**a**) QSY-21. Error bars show the standard deviation of three samples and are smaller than the data point. (**b**) TB. Error bars show the standard deviation of four samples

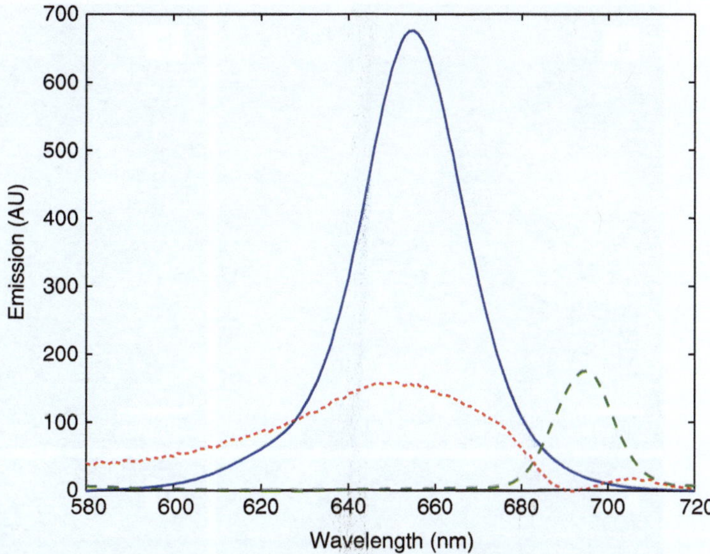

Fig. 3 Emission from 4 nM QDs (*blue, solid*), 10 μM QSY-21 (*green, dashed*, 10× scale), and 10 μM TB (*red, dotted*, 100× scale) in PBS

For this application it is commonly used at millimolar concentrations, further suggesting that cytotoxicity is negligible.

3.2.3 Verification of QD Internalization and Quenching

1. Incubate cells with QDs for either binding (4 °C, 10 min) or internalization (37 °C, 1 h).
2. Aspirate cell culture medium and wash with imaging medium.
3. Add 1 mL of imaging medium and quencher to desired concentration. For the images in Fig. 4, 1 μM QSY-21 was used.

For cold-bound QDs on the plasma membrane, the addition of quencher eliminates the fluorescent signal (Fig. 4a, b). If the QDs are internalized by the cell, the addition of quencher eliminates the out-of-plane fluorescence from the QDs that remain on the cell surface, leading to a more clear image, while QDs inside the cell remain visible (Fig. 4c, d). This is accomplished without the use of a pinhole to exclude out-of-plane fluorescence allowing full-frame images to be acquired in real time.

3.3 Transmission Electron Microscopy

The electron-dense core of QDs offers the possibility of using QDs as probes for correlative electron and fluorescence imaging. Additionally, the routine sectioning of cells for TEM provides a straightforward approach to distinguish intracellular QDs from plasma membrane-bound QDs. However the relatively low electron density of QDs, coupled with the presence of relatively electron-dense regions of the cell following fixation and staining, has made it difficult to image QDs with TEM [7, 8].

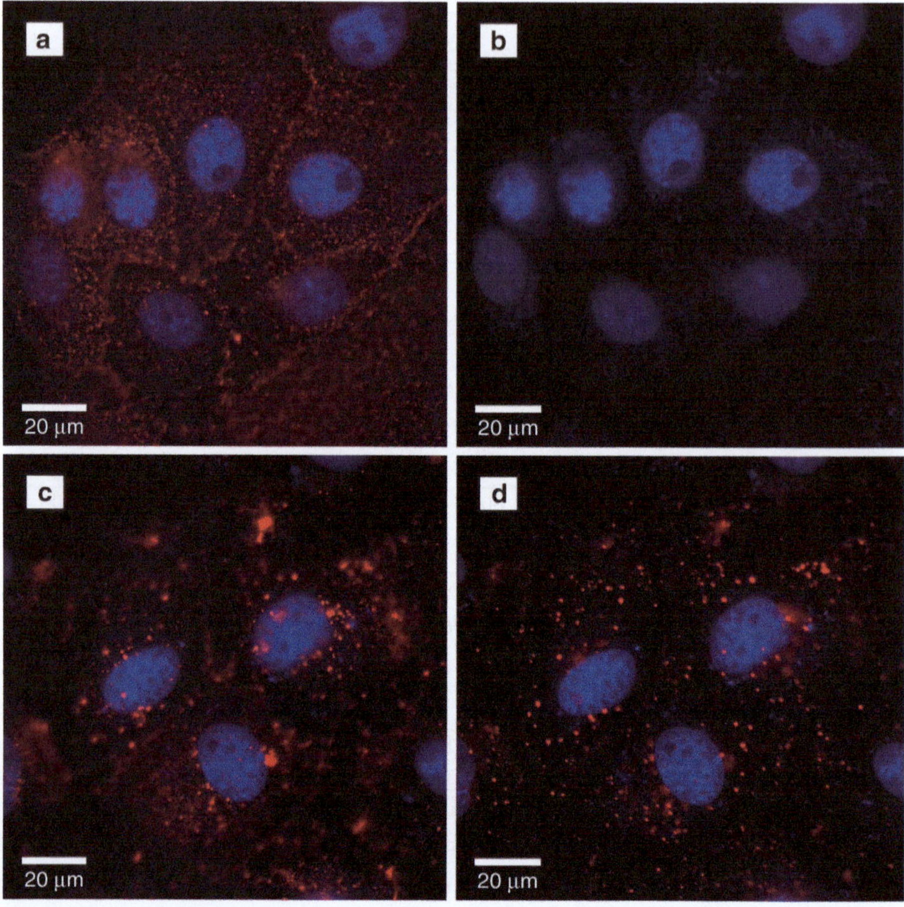

Fig. 4 (**a**) QDs (*red*) bound to the plasma membrane of BS-C-1 cells. Incubation at 4 °C prevents internalization. Nuclei are stained with DAPI (*blue*). (**b**) The addition of 1 μM QSY-21 quenches the QD signal. (**c**) QDs incubated with BS-C-1 cells at 37 °C for 1 h. (**d**) The addition of 1 μM QSY-21 does not affect QDs within the cell, but does quench fluorescence from QDs remaining on the plasma membrane

To obtain high-quality TEM images of QDs, we have extended the use of a standard silver enhancement method for use with QDs. Silver enhancement has been widely used to increase the size of gold particles used in immunogold labeling [28–33]. In this technique, metallic silver derived from reduced ionic silver nucleates around the gold particles in the presence of a reducing agent. This increases the particle size such that subnanometer gold particles can be easily visualized on a TEM. In the published literature describing the silver enhancement technique, silver lactate is often used as a source of silver ions. Its low dissociation constant allows it to be reduced more evenly in the sample. Hydroquinone is often used as a reducing agent. A protective colloid can also be used to inhibit the reaction between the silver ions and the reducing agent. The silver enhancement technique is here employed to coat QDs for imaging with conventional bright-field TEM imaging (Figs. 5 and 6).

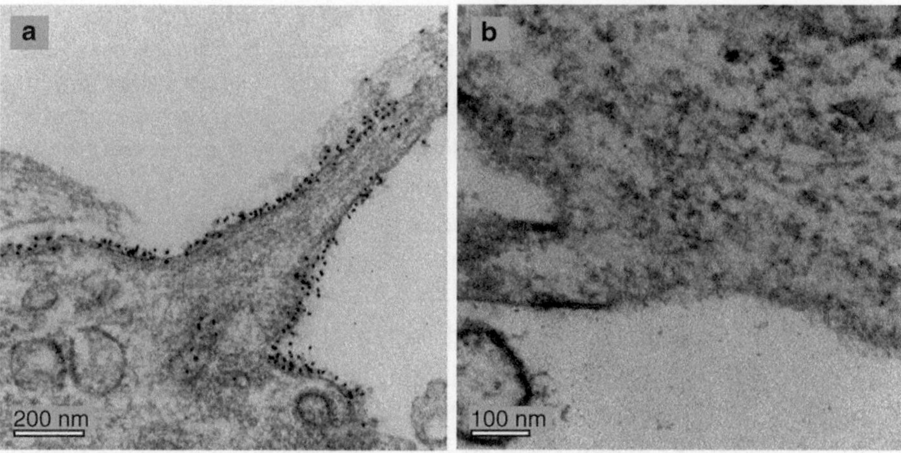

Fig. 5 TEM images of QDs on the cell membrane. (**a**) Silver enhancement, (**b**) control in the absence of silver enhancement. Surface binding of QDs was confirmed by fluorescence imaging of fixed cells with widefield fluorescence microscopy prior to preparation for TEM

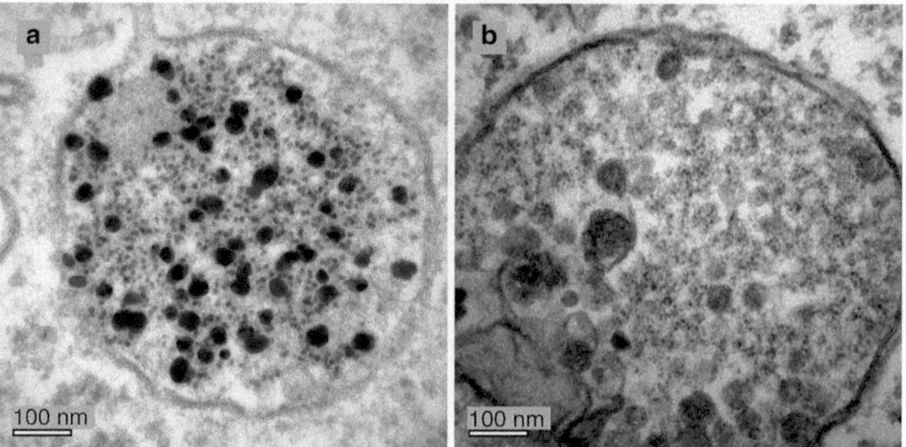

Fig. 6 QDs localized in endocytic vesicles. (**a**) Silver enhancement. The endocytic vesicle contains a combination of individual and aggregated QDs. (**b**) Control in the absence of silver enhancement

3.3.1 Binding and Internalization of Quantum Dots

See Subheading 3.1.

3.3.2 Preparation of Cells for TEM

This protocol describes volumes used for 3.5 cm cell culture dishes and can be scaled as necessary.

1. Fix cells by adding 1–2 mL of the fixative to the cell culture dish at room temperature. Store at 4 °C for at least 2 h, until ready for **step 2**.
2. Remove fixative and rinse with rinse buffer 2 times for 2 min.
3. Remove rinse buffer and rinse with ECS 4 times for 2 min.

4. Prepare silver enhancement solution according to the manufacturer's instructions immediately prior to use. Remove ECS and apply 1 mL of freshly mixed silver enhancement solution to the cell culture dish, cover the dish with aluminum foil, place the dish on a shaker, and shake for 1 h at room temperature.
5. Terminate the enhancement process by removing the enhancement solution from the dish and adding 1 ml enhancement termination solution. Continue shaking for 5 min followed by three rinses with ECS for 2 min each.
6. Remove ECS and rinse with rinse buffer three times for 2 min.
7. Postfix cells by adding 1 mL of post-fixative and let sit for 15 min at room temperature.
8. Remove post-fixative solution and briefly rinse with deionized water.
9. Dehydrate the cells through a series of rinses with increasing concentrations of ethanol in water. For each rinse, add 2 mL of ethanol solution and shake for 2 min. Successive rinse concentrations are 25, 50, 70, 95 %, and 3× 100 % ethanol in water. Perform one final rinse with fresh 100 % ethanol for 2 min while shaking.
10. Remove the final ethanol solution. Add a freshly prepared solution of resin-absolute ethanol (50/50 by volume) to the cells and incubate for 2 h at room temperature.
11. Replace the resin-ethanol solution with 100 % resin and incubate for 8 h. Replace with 100 % resin until ready for resin curing.
12. When ready for curing, fill a plastic capsule with resin and invert it onto the cells.
13. Cure the resin-embedded cells in an oven at 60 °C for 48 h.

3.3.3 Cell Sectioning and Staining

Remove the resin block with the embedded cells from the culture dish, and saw the block into the desired sizes. Glue the small blocks onto a resin stub with the cell side facing up, so cells can be sectioned horizontally. Cut sections with a diamond knife on an ultramicrotome at a 60–70 nm thickness and collect the sections onto 200 mesh copper grids. After sectioning the cells and collecting the sections onto copper grids, counterstain the sections with 5 % uranyl acetate for 5 min. Rinse thoroughly with double-distilled water, followed by 2 % lead citrate for 10 min. Rinse again thoroughly with double-distilled water.

3.3.4 Transmission Electron Microscopy

TEM images (Figs. 5 and 6) were acquired under a 75 kV accelerated voltage and a 20 μm objective aperture on a Hitachi H-7500 TEM. Figure 5 shows a comparison of plasma membrane-bound QDs with (Fig. 5a) and without (Fig. 5b) silver enhancement.

As described in Subheading 3.2.3, **steps 1** and **2**, cells were incubated with QDs at 4 °C to prevent internalization. The silver-enhanced QDs are visible as distinct, individual QDs on the plasma membrane. In comparison, no QDs are visible in the absence of silver enhancement. Both cells were incubated with the same concentration (4 nM) of QDs from the same sample. The cells were treated identically with the exception of the silver enhancement. Additionally, the cells were imaged with fluorescence microscopy after fixation, but before further TEM preparation steps, to confirm QD binding (data not shown). Figure 6 shows QDs localized in endocytic vesicles with (Fig. 6a) and without (Fig. 6b) silver enhancement. QDs are highly concentrated and visible in both vesicles, but have much higher contrast in the silver-enhanced image. The QDs are largely visible in the non-enhanced image only where QDs overlap. QDs in the non-enhanced image have an average size of 9.1 × 6.4 nm while QDs in the enhanced image have an average size of 10 × 8.6 nm. The much larger, darker features in the enhanced image contain several QDs and likely aggregated silver.

4 Note

1. The cost of the quencher can also be considered. At the time of publication, the cost of 1 mL of 1 μM solution (one cell plate from the assay described in this chapter) of QSY-21 is \$0.031, as compared to TB for which the cost is \$$1.1 \times 10^{-6}$ (purchased as a solid). This factor of ~28,000 in cost may be trivial for some experiments, but for large-scale applications of this method, such as flow cytometry, it may be a considerable expense.

Acknowledgments

C.K.P. gratefully acknowledges financial support from NIH R01-GM086195.

References

1. Chan WC, Nie S (1998) Quantum dot bioconjugates for ultrasensitive nonisotopic detection. Science 281:2016–2018
2. Michalet X, Pinaud F, Bentolila L, Tsay J, Doose S, Li J, Sundaresan G, Wu A, Gambhir S, Weiss S (2005) Quantum dots for live cells, in vivo imaging, and diagnostics. Science 307:538–544
3. Alivisatos AP, Gu W, Larabell C (2005) Quantum dots as cellular probes. Annu Rev Biomed Eng 7:55–76
4. Medintz IL, Uyeda HT, Goldman ER, Mattoussi H (2005) Quantum dot bioconjugates for imaging, labelling and sensing. Nat Mater 4:435–446

5. Gao X, Cui Y, Levenson RM, Chung LWK, Nie S (2004) In vivo cancer targeting and imaging with semiconductor quantum dots. Nat Biotechnol 22:969–976
6. Chan W, Maxwell D, Gao X, Bailey R, Han M, Nie S (2002) Luminescent quantum dots for multiplexed biological detection and imaging. Curr Opin Biotech 13:40–46
7. Lagerholm BC, Wang M, Ernst LA, Ly DH, Liu H, Bruchez MP, Waggoner AS (2004) Multicolor coding of cells with cationic peptide coated quantum dots. Nano Lett 4:2019–2022
8. Nisman R, Dellaire G, Ren Y, Li R, Bazett-Jones DP (2004) Application of quantum dots as probes for correlative fluorescence, conventional, and energy-filtered transmission electron microscopy. J Histochem Cytochem 52: 13–18
9. Tsien RY (1998) The green fluorescent protein. Annu Rev Biochem 67:509–544
10. Pinaud F, Clarke S, Sittner A, Dahan M (2010) Probing cellular events, one quantum dot at a time. Nat Methods 7:275–285
11. Delehanty JB, Bradburne CE, Boeneman K, Susumu K, Farrell D, Mei BC, Blanco-Canosa JB, Dawson G, Dawson PE, Mattoussi H, Medintz IL (2010) Delivering quantum dot-peptide bioconjugates to the cellular cytosol: escaping from the endolysosomal system. Integr Biol 2:265–277
12. Frasco MF, Chaniotakis N (2009) Bioconjugated quantum dots as fluorescent probes for bioanalytical applications. Anal Bioanal Chem 396:229–240
13. Pawley J (2006) Handbook of biological confocal microscopy, 3rd edn. Springer, New York, NY
14. Kameyama S, Horie M, Kikuchi T, Omura T, Tadokoro A, Takeuchi T, Nakase I, Sugiura Y, Futaki S (2007) Acid wash in determining cellular uptake of Fab/cell-permeating peptide conjugates. Pept Sci 88:98–107
15. Haigler HT, Maxfield FR, Willingham MC, Pastan I (1980) Dansylcadaverine inhibits internalization of 125I-epidermal growth factor in BALB 3 T3 cells. J Biol Chem 255:1239–1241
16. Kamen BA, Wang MT, Streckfuss AJ, Peryea X, Anderson RG (1988) Delivery of folates to the cytoplasm of MA104 cells is mediated by a surface membrane receptor that recycles. J Biol Chem 263:13602–13609
17. Leamon CP, Low PS (1993) Membrane folate-binding proteins are responsible for folate-protein conjugate endocytosis into cultured cells. Biochem J 291:855–860
18. Shoji Y, Akhtar S, Periasamy A, Herman B, Juliano RL (1991) Mechanism of cellular uptake of modified oligodeoxynucleotides containing methylphosphonate linkages. Nucl Acids Res 19:5543–5550
19. Xu C, Xing B, Rao J (2006) A self-assembled quantum dot probe for detecting β-lactamase activity. Biochem Bioph Res Co 344:931–935
20. Tomasulo M, Yildiz I, Raymo FM (2006) pH-Sensitive quantum dots. J Phys Chem B 110: 3853–3855
21. Chen C, Cheng C, Lai C, Wu P, Wu K, Chou P, Chou Y, Chiu H (2006) Potassium ion recognition by 15-crown-5 functionalized CdSe/ZnS quantum dots in H_2O. Chem Commun 263–265
22. Dyadyusha L, Yin H, Jaiswal S, Brown T, Baumberg JJ, Booy FP, Melvin T (2005) Quenching of CdSe quantum dot emission, a new approach for biosensing. Chem Commun 3201–3203
23. Somers RC, Bawendi MG, Nocera DG (2007) CdSe nanocrystal based chem-/bio- sensors. Chem Soc Rev 36:579–591
24. Oh E, Hong M, Lee D, Nam S, Yoon HC, Kim H (2005) Inhibition assay of biomolecules based on fluorescence resonance energy transfer (FRET) between quantum dots and gold nanoparticles. J Am Chem Soc 127: 3270–3271
25. Howarth M, Liu W, Puthenveetil S, Zheng Y, Marshall LF, Schmidt MM, Wittrup KD, Bawendi MG, Ting AY (2008) Monovalent, reduced-size quantum dots for imaging receptors on living cells. Nat Methods 5:397–399
26. Jablonski AE, Kawakami T, Ting AY, Payne CK (2010) Pyrenebutyrate leads to cellular binding, not intracellular delivery, of polyarginine quantum dots. J Phys Chem Lett 1: 1312–1315
27. Payne CK, Jones SA, Chen C, Zhuang X (2007) Internalization and trafficking of cell surface proteoglycans and proteoglycan-binding ligands. Traffic 8:389–401
28. Danscher G (1981) Histochemical demonstration of heavy metals. Histochemistry 71:1–16
29. Danscher G, Norgaard J (1983) Light microscopic visualization of colloidal gold on resin-embedded tissue. J Histochem Cytochem 31: 1394–1398
30. Danscher G, Stoltenberg M (2006) Silver enhancement of quantum dots resulting from (1) metabolism of toxic metals in animals and humans, (2) in vivo, in vitro and immersion created zinc–sulphur/zinc–selenium nanocrystals, (3) metal ions liberated from metal implants and particles. Prog Histochem Cyto 41:57–139

31. Chou LYT, Fischer HC, Perrault SD, Chan WCW (2009) Visualizing quantum dots in biological samples using silver staining. Anal Chem 81:4560–4565
32. Burry R, Vandre D, Hayes D (1992) Silver enhancement of gold antibody probes in pre-embedding electron microscopic immunocytochemistry. J Histochem Cytochem 40:1849–1856
33. Javois L (1999) Immunocytochemical methods and protocols, 2nd edn. Humana Press, Totowa, NJ

Chapter 3

Imaging of Cell Populations in Atherosclerosis Using Quantum Dot Nanocrystals

Joshua R. Trantum and Ashwath Jayagopal

Abstract

Atherosclerosis, a leading cause of morbidity and mortality worldwide, is characterized by the accumulation of lipid deposits inside arterial walls, leading to narrowing of the arterial lumen. A significant challenge in the development of diagnostic and therapeutic strategies is to elucidate the contribution of the various cellular participants, including macrophages, endothelial cells, and smooth muscle cells, in the initiation and progression of the atheroma. This protocol details a strategy using quantum dot nanocrystals to monitor homing and distribution of cell populations within atherosclerotic lesions with high signal to noise ratios over prolonged periods of analysis. This fluorescence-based approach enables the loading of quantum dots into cells such as macrophages without perturbing native cell functions in vivo, and has been used for the multiplexed imaging of quantum dot-labeled cells with biomarkers of atherosclerotic disease using conventional immunofluorescence techniques.

Key words Nanotechnology, Quantum dots, Nanocrystals, Macrophages, Atherosclerosis, Vascular biology, Fluorescence imaging, Leukocytes

1 Introduction

A major objective in efforts to achieve early detection and effective treatment of atherosclerotic disease is to identify the major cellular and molecular mediators of the disease [1, 2]. For achieving this goal, nanotechnology offers promising approaches for targeting imaging agents to specific biomolecules. Nanoparticles may be used to identify important constituents of the atherosclerotic plaque, including endothelial cells, macrophages, and lymphocytes [1, 3]. An emerging challenge is to develop imaging strategies for targeting imaging agents toward multiple cell types within the plaque such that multiplexed imaging of cellular components may be simultaneously imaged.

Toward the goal of enabling detection of cell subpopulations and analysis of their respective roles in plaque biology, we describe a

Sandra J. Rosenthal and David W. Wright (eds.), *NanoBiotechnology Protocols*, Methods in Molecular Biology, vol. 1026, DOI 10.1007/978-1-62703-468-5_3, © Springer Science+Business Media New York 2013

preclinical nanotechnology-based optical imaging approach [4]. A technique based on this approach is described which utilizes fluorescent quantum dot (QD) nanocrystals for the ex vivo labeling of macrophages, and the analysis of QD-labeled cell distribution within atherosclerotic plaques of the ApoE-deficient mouse model of atherosclerosis (ApoE $^{-/-}$). QD play an important role in this approach due to their size-tunable distinct emission profiles to color-code distinct cell populations, their high quantum efficiency, amenability to conjugation with targeting ligands, and resistance to photobleaching, all of which enable multiplexed long-term imaging of QD-labeled cells and disease biomarkers within the atheroma [5–8]. In this protocol, QD are linked to the cell-penetrating peptide maurocalcine [9] to enable loading of live macrophages with nanocrystals without toxicity or adverse effects on cell function. Techniques for monitoring the efficiency of QD loading into cells and the ex vivo analysis of QD-labeled macrophages within atherosclerotic plaques are outlined. While the macrophage is the cell target of interest in this protocol, the techniques herein can be applied toward ex vivo optical imaging of other cell types, such as T lymphocytes within the atherosclerotic plaque, with similar effectiveness [4].

2 Materials

1. Sterile distilled deionized water (Millipore, Inc.).
2. Maurocalcine, 33-amino-acid custom-synthesized peptide (several custom peptide synthesis services, such as Biomatik, Inc. and Genscript, Inc., can be consulted to synthesize the maurocalcine peptide). Details of this peptide's synthesis have been described elsewhere [9]. Specifically, the peptide was synthesized and purified such that the product was 98 % pure, as validated by HPLC and MS verification. The N-terminus was biotinylated to enable its conjugation to streptavidin-functionalized quantum dots, and the C-terminus was amidated to avoid nonspecific peptide interactions. Additionally, the vendor aliquoted the peptide into single-use 1 mg aliquots to facilitate easy dispensing of the lyophilized peptide without unnecessary freeze-thaw cycles. Upon receipt of lyophilized peptides, aliquots were stored at −20 °C. *Important*: The potential toxicity of this peptide has not been extensively evaluated, and therefore, handling of this peptide should be carried out according to established safety procedures for biohazardous, nonvolatile substances.
3. 655 nm-emitting ITK QD-Streptavidin Conjugate Kit, 2 μM solution (Cat # Q10021MP, Invitrogen Corp.), featuring 6–8

streptavidin proteins per QD. This is referred to as "QD-SAV" in Subheading 3.

4. Borate buffered saline, pH = 8.2 (Cat # 08059, Sigma-Aldrich, Inc.).
5. Eppendorf Protein LoBind 1.5 mL tubes (Cat # 0030 108.116, Eppendorf North America, Inc.).
6. Amicon Ultra-4 50 K MWCO spin column centrifugation devices (Cat # UFC805008, Millipore, Inc.).
7. Swinging bucket centrifuge (Allegra X-22R, Beckman Coulter, Inc.).
8. AutoMACS rinsing solution of PBS with 2 mM EDTA adjusted to pH = 7.2 (Cat # 130-091-222, Miltenyi Biotec), diluted with MACS BSA stock solution (Cat # 130-091-376, Miltenyi Biotec) to achieve 0.5 % BSA concentration.
9. C57/BL6 mice, aged 8–12 weeks (Harlan, Inc.) [used for isolation of mouse macrophages from spleen].
10. Isoflurane anesthesia with vaporizer (Terrell).
11. 15 mL sterile conical tubes (Cat # 07-200-886, Fisher Scientific, Inc.).
12. 5 mL syringes (Cat # 305218, BD Biosciences) [plunger used in isolation procedure].
13. Cell strainer with 70 μm nylon mesh (Cat # 352350, BD Biosciences).
14. INCYTO C-Chip Hemacytometer (Cat # 22-600-101, Fisher Scientific, Inc.).
15. Trypan Blue (Cat # SV30084.01, Thermo Scientific, Inc.).
16. CD11b positive selection magnetic microbeads (Cat # 130-049-601, Miltenyi Biotec) [one kit is used for isolation of up to 1×10^9 CD11b$^+$ mouse cells].
17. MiniMACS Separator Unit with MS magnetic enrichment columns (Cat # 130-090-312, Miltenyi Biotec).
18. Thermo Scientific Barnstead LabQuake rotator (Cat # 4002110Q, Thermo Scientific).
19. LSR II flow cytometer with UV diode (405 nm) and argon (488 nm) laser or equivalent excitation sources (BD Biosciences).
20. Anti-mouse CD68 antibody, Alexa Fluor 488 conjugate (Cat # MCA1957A488, AbD Serotec, Inc.), to identify macrophages.
21. Physiological saline (Cat # S77939, Fisher Scientific, Inc.).
22. 7-month-old ApoE$^{-/-}$ mice (Harlan, Inc.).

23. Phosphate buffered saline (Cat # BP2438-4, Fisher Scientific, Inc.).
24. 10 % neutral buffered formalin (Cat # 22-046-361, Fisher Scientific, Inc.).
25. Liquid nitrogen.
26. Dry ice.
27. OCT compound (Cat # 14-373-65, Fisher Scientific, Inc.).
28. TE2000U inverted fluorescence microscope (Nikon Instruments, Inc.).
29. C7780 color CCD camera (Hamamatsu).
30. Image Pro Plus image processing software (Media Cybernetics, Inc.).
31. ProLong Gold antifade mounting medium (Cat # P36930, Invitrogen Corp.).

3 Methods

3.1 Functionalization of QD with the Cell-Penetrating Peptide Maurocalcine

1. Resuspend 1 mg of lyophilized maurocalcine in 1 mL of sterile distilled water (= 245 μM stock solution), and incubate at room temperature, sealed, for 2 h. This will ensure complete dissolution of the peptide (*see* **Note 1**). Any unused stock solution may be frozen at −80 °C for long-term storage and reused no more than twice for QD conjugations; therefore, it is important to aliquot the peptide according to its anticipated number of conjugations that will be required.
2. For every 1×10^6 macrophages to be labeled with QD, prepare an Eppendorf tube with 12.5 μL QD-SAV and 3.28 μL maurocalcine in 184.22 μL of borate-buffered saline. This constitutes a labeling solution with 125 nM QD-SAV and 4,000 nM maurocalcine. Allow this solution to incubate at room temperature for 30 min to allow streptavidin-biotin binding interactions. The concentrations of each component in the labeling solution are utilized to ensure that maurocalcine saturates the surface of the QD-SAV, as there are 6–8 streptavidins per QD, with each streptavidin having 4 biotin-binding sites, and therefore, 32 available maurocalcine-binding sites per QD-SAV.
3. Load each sample onto an Amicon Ultra-4 spin column device, using gentle pipetting along the area of the UltraCel membrane surface. Gently add to this solution 3.8 mL of AutoMACS rinsing solution.
4. In a swinging bucket centrifuge, centrifuge the devices at $4{,}000 \times g$ for 15 min at room temperature. This step serves to remove excess, unbound maurocalcine from the QD-SAV/

maurocalcine solution using size-exclusion spin chromatography.

5. Recover the retentate, which is approximately 80 μL, using a pipettor, and transfer to 1.5 mL tubes. Retain at room temperature in the dark until ready for use (*see* **Note 2**). Discard the filtrate containing unbound maurocalcine as biohazardous waste.

3.2 Immunomagnetic Isolation of Mouse Macrophages

It is important in this procedure to work on ice using the recommended rinsing solution, to ensure preservation of CD11b antigens used for isolation, and to prevent cell aggregation and nonspecific binding to beads.

1. Anesthesize mice with isoflurane (4 % induction) and euthanize by cervical dislocation.
2. Excise spleens from mice and immediately place each spleen in 15 mL conical tubes filled with AutoMACS solution with BSA chilled to 4 °C ("AutoMACS") (*see* **Note 3**).
3. On ice, disrupt cells with the plunger of a 5 mL syringe until a turbid cell suspension is obtained. Titurate the suspension with a 10 mL pipet, and strain the cells three times through a 70 μm cell strainer.
4. Centrifuge the cell suspension at 450 × *g* for 5 min at 4 °C, and aspirate the supernatant. Resuspend the cells with 1 mL of AutoMACS.
5. Using a hemacytometer, perform a total cell count. Use trypan blue exclusion to determine dead cell fraction and discard from total cell count in order to calculate the live cell concentration. In our experience, viability is >90 % at this point in the procedure, and the total cell number should be between 1 and 2×10^8 cells per spleen, depending on the efficiency of the tissue disruption procedure. A cell count significantly below this amount will indicate that the splenic disruption was incomplete and may likely need to be repeated. Keep cells on ice.
6. Resuspend the cells in 1 mL of buffer for every 1×10^8 cells. Add 40 μL of $CD11b^+$ microbead solution for every 1×10^8 cells, and incubate for 30 min on ice, gently inverting the tube every 5 min to ensure proper mixing.
7. Rinse the cells by centrifugation at 450 × *g* for 5 min at 4 °C, aspirating supernatant and resuspending the cell pellet in 1 mL AutoMACS. Repeat this centrifugation and wash step twice, but in the last resuspension step, resuspend cells in 500 μL of AutoMACS.
8. Place MS column within the MiniMACS separator by fitting the column in the designated slot. Rinse the column with 500 μL of AutoMACS.

9. Carefully overlay the 500 μL cell suspension onto the MS column.
10. Collect 3×500 μL fractions of unlabeled cells (CD11b$^-$) by rinsing the MS column with 500 μL of AutoMACS each time an entire column volume is collected. These unlabeled cells can be retained at 4 °C for quality control purposes in Subheading 3.4.
11. To recover the CD11b$^+$ cell fraction containing macrophages, remove the MS column from the MiniMACS device and place it on an Eppendorf tube for collection. Use 1 mL of AutoMACS solution and the supplied plunger to gently expel the CD11b$^+$ cells from the column. Retain collected cells for labeling with QD in Subheading 3.3. Discard columns as biological waste after rinsing with 5 mL of household bleach. A fraction of the collected cells should be counted using a hemacytometer with trypan blue exclusion to determine the yield of macrophages, which is typically 2×10^6 cells per spleen.

3.3 Intracellular Loading of Macrophages with QD

1. In 15 mL sterile conical tubes, combine collected cells from Subheading 3.2, **step 11**, with QD-SAV/maurocalcine labeling solutions from Subheading 3.1, **step 5**. Next, add 10 μL of the anti-CD68 antibody solution, and dilute the solution to 2.5 mL using AutoMACS. Incubate the solution on a tube rotator in a humidified cell culture incubator at 37 °C, 5 % CO_2 for 30 min (*see* **Note 4**).
2. Centrifuge the cells at 450×*g*, aspirate the supernatant without disturbing the cell pellet, and rinse with 1 mL of AutoMACS. Repeat thrice, resuspending the cells in the final step to 500 μL of AutoMACS.
3. Prepare to inject cells in animal models immediately for optimal homing to atherosclerotic lesions (Subheading 3.5.1), while concurrently having a fraction of the cells analyzed by flow cytometry (Subheading 3.4). It is recommended that each task be delegated to different personnel to minimize the time between cell labeling and injection procedures.

3.4 Monitoring of Loading Efficiency

1. Dilute a 50 μL fraction of the labeled cells from **step 2** of Subheading 3.3 with 1 mL of AutoMACS, and analyze labeling efficiency on a flow cytometer configured for Alexa Fluor 488 excitation and emission settings (488 nm excitation, 520/20 nm emission) to visualize CD11b$^+$ cells, and QD excitation and emission settings (405 nm excitation with 655/20 nm emission) to visualize the fraction of those cells which are QD$^+$. In our hands, the labeling procedure achieved >90 % labeling efficiency of CD68$^+$ macrophages, and very few contaminating cell types, such as lymphocytes, were present in the purified cell mixture [4]. As a negative control, the CD11b$^-$ cell population from the elution steps of Subheading 3.2, **step 10**, can be

analyzed using these settings following dilution in AutoMACS and CD68 staining to validate the purity of the procedure, and to set background settings for measuring QD fluorescence.

2. Use flow cytometry analysis software (e.g., FlowJo, Treestar, Inc.) to analyze the percentage of $QD^+/CD68^+$ macrophages, using the region of interest function.

3.5 Analysis of QD-Loaded Macrophages in the ApoE $^{-/-}$ Mouse Model of Atherosclerosis

In our experience, injection of 1×10^6 QD-labeled cells is sufficient to visualize macrophages within atherosclerotic plaques following only 2 days of circulation time. However, cells can be visualized as long as 1 month following injection of cells, and injected doses may be varied from 1 to 5×10^6 cells with similar results.

3.5.1 Injection of Cells

1. Centrifuge 1×10^6 cells at $450 \times g$ and resuspend in 100 μL of sterile saline.
2. Inject cells into retro-orbital plexus of ApoE$^{-/-}$ mice.
3. Immediately following injection, apply firm pressure on the injected eye using gauze pads to encourage hemostasis.

3.5.2 Tissue Collection and Processing

Following a circulation period of 2 days to 1 month, euthanize mice as described in Subheading 3.2.

1. Harvest aortas of mice, which contain substantial aortic plaques at 7 months of age (*see* **Note 5**).
2. Rinse aorta with PBS chilled to 4 °C to remove blood and debris.
3. Place tissue into neutral-buffered formalin and fix overnight at room temperature.
4. Rinse in 4 °C PBS three times for 10 min each. During this step, a stainless steel container should be filled with liquid nitrogen and another container should be filled with dry ice. Partially fill the mold with OCT.
5. Transfer aorta to a clean petri dish and absorb excess PBS with a clean tissue.
6. Place tissue in pre-labeled base molds filled with OCT. Arrange the tissue within the OCT as flat as possible, and near the bottom of the space, to facilitate cryosectioning.
7. Use a forceps to hold base mold edge and place the base mold into the surface of the liquid nitrogen. Allow bottom of mold to touch the surface of the liquid nitrogen. Hold until the aorta solidifies. Immediately remove tissue once solidified to avoid cracking of the mold.
8. Place tissue block on dry ice.
9. Store frozen tissue block in −80 °C freezer immediately and maintain until sectioning.
10. For sectioning, attach the frozen tissue block on the cryostat chuck. Allow tissue block to equilibrate to the cryostat temperature

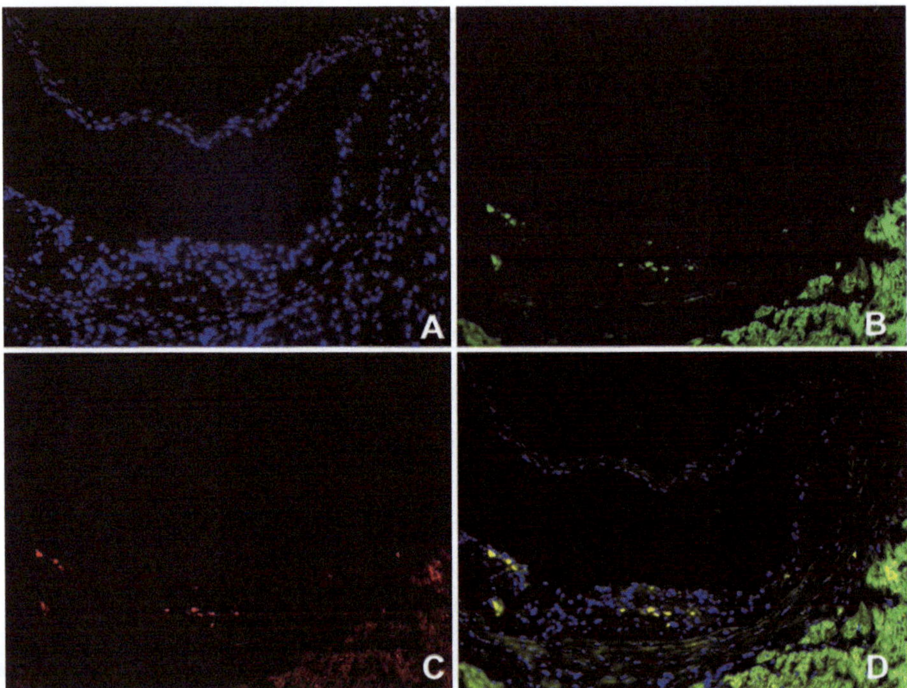

Fig. 1 Visualization of QD-labeled macrophages within atherosclerotic plaques within the aortic root. (**a**) DAPI nuclear counterstaining; (**b**) Alexa Fluor anti-CD68 staining, to identify macrophages; (**c**) QD-labeled macrophages; (**d**) overlaid image; yellow regions correspond to CD68$^+$/QD$^+$ macrophages. Elastin accounts for the intense hyperfluorescence in all channels in the lower right of the image, and should be expected as part of immunofluorescence analysis

(−20 °C) before cutting sections. For our analyses, sections are cut at 5 μm. It is recommended that a trained histologist carry out these steps (*see* **Note 6**).

11. Dry at room temperature till the sections are firmly adhered to the slide, and mount slide in ProLong Gold mounting media with DAPI and cure medium to slide by overnight storage at room temperature. Store in a dry, dark location until image analysis. It is recommended that slides be imaged within the week following sectioning to maximize QD fluorescence.

3.5.3 Tissue Imaging

For imaging of QD in aortic atherosclerotic plaques, we used a Nikon TE2000U inverted microscope with Hamamatsu C7780 CCD camera, in conjunction with Image Pro Plus 5.1 for image acquisition. However, any microscope configuration designed for fluorescence imaging using the excitation/emission settings described below may be appropriate.

1. Locate aortic root lesions in the ascending aorta through light micrographic imaging of 5 μm sections (Fig. 1).

2. In fluorescence mode, acquire images using DAPI (360/40 nm excitation, 460/20 nm emission) and QD (405/40 nm excitation, 655/20 nm emission) settings.
3. Use the "color composite" function of Image Pro Plus to overlay nuclear images (DAPI) with QD and light micrographs as desired to photograph plaques. Colocalization analysis can be performed at this point to enumerate QD^+ nuclei within the plaque, representing the injected macrophages.

4 Notes

1. It is important to dissolve the peptide in water, as salt solutions impede optimal dissolution of the peptide. This stock solution can be diluted with saline, however, without adverse effects on the peptide.
2. The QD-SAV/maurocalcine can be stored in this manner for same-day use. For use up to 3 days later, the conjugate should be stored at 4 °C in the dark.
3. One homogenized spleen yields approximately 10×10^8 total cells, of which 5 % of the population consists of $CD11b^+$ monocyte/macrophage cells. The immunomagnetic isolation procedure, in our hands, provided approximately a 40 % yield of $CD11b^+$ cells. Therefore, as a guideline for planning isolations, using the immunomagnetic isolation procedure one can expect approximately 2×10^6 $CD11b^+$ macrophages per spleen to be recovered for use in subsequent QD labeling and imaging experiments.
4. This step can accommodate from 0.5 to 2×10^6 cells from **step 11** of Subheading 3.2, with similar labeling efficiencies. Other cell concentration ranges have not been tested.
5. While $ApoE^{-/-}$ mice fed on chow diet at 7 months of age do feature aortic plaques, larger plaque regions may be obtained by feeding mice on a high-fat diet and/or using older (12 months) ApoE-deficient mice.
6. Cryosectioning is best carried out by a histologist. Consulting with a histologist prior to sectioning tissues can also enable a number of specialized techniques, such as those using oil red O staining to visualize lipid within plaques, CD68 staining to visualize macrophages, and CD4/8 staining to visualize lymphocytes. The distinct emission spectra of QD enable multiplexed imaging of cell populations within the plaque, as previously described in ref. [4].

Acknowledgments

This work was supported in part by a National Eye Institute Core Grant in Vision Research (P30-EY008126), and the International Retinal Research Foundation.

References

1. Jayagopal A, Linton MF, Fazio S, Haselton FR (2010) Insights into atherosclerosis using nanotechnology. Curr Atheroscler Rep 12:209–215
2. Libby P, DiCarli M, Weissleder R (2010) The vascular biology of atherosclerosis and imaging targets. J Nucl Med 51(Suppl 1):33S–37S
3. Sadeghi MM, Glover DK, Lanza GM, Fayad ZA, Johnson LL (2010) Imaging atherosclerosis and vulnerable plaque. J Nucl Med 51(Suppl 1): 51S–65S
4. Jayagopal A, Su YR, Blakemore JL, Linton MF, Fazio S, Haselton FR (2009) Quantum dot mediated imaging of atherosclerosis. Nanotechnology 20:165102
5. Jayagopal A, Russ PK, Haselton FR (2007) Surface engineering of quantum dots for in vivo vascular imaging. Bioconjug Chem 18: 1424–1433
6. Rosenthal SJ, Tomlinson I, Adkins EM et al (2002) Targeting cell surface receptors with ligand-conjugated nanocrystals. J Am Chem Soc 124:4586–4594
7. Smith AM, Gao X, Nie S (2004) Quantum dot nanocrystals for in vivo molecular and cellular imaging. Photochem Photobiol 80:377–385
8. Gao X, Chung LW, Nie S (2007) Quantum dots for in vivo molecular and cellular imaging. Methods Mol Biol 374:135–145
9. Esteve E, Mabrouk K, Dupuis A et al (2005) Transduction of the scorpion toxin maurocalcine into cells. Evidence that the toxin crosses the plasma membrane. J Biol Chem 280: 12833–12839

Chapter 4

Imaging of Endothelial Progenitor Cell Subpopulations in Angiogenesis Using Quantum Dot Nanocrystals

Joshua M. Barnett, John S. Penn, and Ashwath Jayagopal

Abstract

Over the last decade, research has identified a class of bone marrow-derived circulating stem cells, termed endothelial progenitor cells (EPCs), that are capable of homing to vascular lesions in the eye and contributing to pathological ocular neovascularization (NV). In preclinical and biological studies, EPCs are frequently identified and tracked using a intracellularly loaded fluorescent tracer, 1,1′-dioctadecyl-3,3,3′,3′-tetramethylindocarbo cyanine perchlorate-labeled acetylated LDL (DiI-acLDL). However, this method is limited by photobleaching and insufficient quantum efficiency for long-term imaging applications. We have developed a method for conjugation of high quantum efficiency, photostable, and multispectral quantum dot nanocrystals (QD) to acLDL for long-term tracking of EPCs with improved signal-to-noise ratios. Specifically, we conjugated QD to acLDL (QD-acLDL) and used this conjugated fluorophore to label a specific $CD34^{+}$ subpopulation of EPCs isolated from rat bone marrow. We then utilized this method to track $CD34^{+}$ EPCs in a rat model of laser-induced choroidal neovascularization (LCNV) to evaluate its potential for tracking EPCs in ocular angiogenesis, a critical pathologic feature of several blinding conditions.

Key words Endothelial progenitor cells, Nanocrystals, Quantum-dots, Choroidal neovascularization, Angiogenesis, Diabetic retinopathy, Age-related macular degeneration, Acetylated low-density lipoprotein, Immunomagnetic cell isolation

1 Introduction

Several highly prevalent blinding ocular conditions are characterized by angiogenesis, including diabetic retinopathy, macular degeneration, and retinopathy of prematurity. A number of cellular and biomolecular interactions contribute to the pathology seen in these disorders [1, 2]. In 1997, Asahara and coworkers identified and initially characterized a circulating population of cells originating from the bone marrow, designated endothelial progenitor cells (EPCs), that were subsequently shown to home to neovascular lesions and contribute to the development of new pathologic vessels [3–6]. Under these conditions, EPCs roll along and adhere to the endothelial cells of the inner vascular wall of neovascular tissue

Sandra J. Rosenthal and David W. Wright (eds.), *NanoBiotechnology Protocols*, Methods in Molecular Biology, vol. 1026,
DOI 10.1007/978-1-62703-468-5_4, © Springer Science+Business Media New York 2013

to augment angiogenesis [7–9]. Several animal models that simulate these ocular diseases have been utilized to image EPCs for elucidation of these and other functions [5, 6]. However, the specific factors governing EPC homing to tissue and the full spectrum of their functions remain incompletely understood. The emerging importance of EPCs in these diseases has motivated the development of methods to image EPCs in tissues with high sensitivity and specificity.

A majority of EPC populations described in the literature originate from the bone marrow and are defined as being CD34 positive ($CD34^+$), and these cells incorporate acetylated low-density lipoprotein (acLDL) [7, 10–15]. Due to the latter characteristic, dye-labeled acLDL incorporation is often used to label and track these cells using the DiI-acLDL fluorophore conjugate (DiI = 1,1′-dioctadecyl-3,3,3′,3′-tetramethylindocarbo-cyanine perchlorate). However, this dye-based strategy is limited by low intensity and photostability, which complicates longitudinal studies of EPCs in disease models. In this chapter, we describe an EPC imaging strategy that incorporates quantum dot nanocrystals (QD) as the fluorophore, rather than DiI. QD are superior to organic dyes in photostability and fluorescence intensity, and thus enable more specific and prolonged visualization of EPCs in tissue. Using QD, EPCs are isolated from rat bone marrow and labeled *ex vivo*. The cells are then systemically injected and imaged in a relevant context using a rat model of laser-induced choroidal neovascularization (LCNV). This animal model is widely employed to model pathologic angiogenesis of the type that occurs in neovascular age-related macular degeneration.

2 Materials

1. Six 4-week-old Brown Norway rats (around 100 g each) [for cell isolation from bone marrow].
2. Isoflurane (Terrell).
3. Hank's Buffered Salt Solution (HBSS): (1× solution): 5 mM KCl, 0.44 mM KH_2PO_4, 137 mM NaCl, 0.34 mM Na_2HPO_4, 3.3 mM $NaHCO_3$, 5.5 mM d-glucose, pH 7.1–7.4 [easily contaminated by bacteria if not kept in a sterile location].
4. Accutase™ (A6964, Sigma).
5. Swinging-bucket centrifuge.
6. Sterile 40 nm nylon mesh (Sefar America, Inc., Fisher Scientific).
7. EasySep® cell isolation system (18558, StemCell Technologies), including: EasySep® FITC Selection Cocktail (18152), magnetic nanoparticles (18150), and EasySep® magnet (18000).

8. Hemacytometer (BD Biosciences, Bedford, MA).
9. Five milliliter Falcon tube: (isolation tube): (352058, Becton Dickinson).
10. Phosphate-buffered saline (PBS): (1× solution): 3.2 mM Na_2HPO_4, 0.5 mM KH_2PO_4, 1.3 mM KCl, 135 mM NaCl, pH 7.4.
11. Isolation buffer: (1× solution): PBS, 2 % fetal bovine serum (FBS), 1.0 mM EDTA [make fresh, important to keep this solution Ca^{2+} and Mg^{2+} free to discourage cell aggregation].
12. FcR blocking antibody (112-001-008, Jackson Immuno Research Laboratories, Inc.).
13. Rabbit anti-CD34 FITC-conjugated antibody (252268, Abbiotec).
14. EGM-2 medium (CC-3162, Clonetics): contains FBS, hydrocortisone, epidermal growth factor, and antibiotic [keeps for a few weeks at 4 °C].
15. 100× antibiotic/antimycotic solution (A5955, Sigma) [keep at −20 °C until needed].
16. Human low-density lipoprotein (Biomedical Technologies, Inc.).
17. Sulfo-NHS-biotin (Pierce protein research products, Thermo Fisher Scientific).
18. Borate buffered saline (BBS): (1× solution): 10 mM sodium borate, 150 mM NaCl, pH 8.2.
19. G-25 Sephadex columns (GE Lifesciences).
20. Sulfo-NHS-acetate (Pierce protein research products, Thermo Fisher Scientific).
21. Sephacryl 400HR (GE Lifesciences).
22. QD655: Quantum dots emitting 655 nm light (Invitrogen).
23. DiI-acLDL (BT-902, Biomedical Technologies, Inc.).
24. 4′,6-diamidino-2-phenylindole (DAPI) (Sigma, D9532).
25. Chamber Slides™ (177429, Lab-Tek).
26. Light microscope similar to the AX70 (Olympus).
27. 100 W mercury lamp and filter cubes (Olympus).
28. Provis system digital camera DP71 (Olympus), computer (Dell), and DP controller software (Olympus).
29. Eight 100 g Brown Norway rats [for induction of LCNV and subsequent analysis of injected QD-labeled EPCs].
30. General anesthetic: 80/12 mg/kg ketamine/xylazine (K113, Sigma).
31. Local anesthetic: 0.5 % proparacaine drops.

32. Dilating eye drops: 2.5 % phenylephrine and 1 % atropine (Bausch and Lomb).
33. 2.5 % Gonak solution (Akorn) [used with glass coverslips to help visualize the fundus of the retina and keep the eye moist during the procedure].
34. Slit lamp (Carl Zeiss Meditec) with a laser delivery system.
35. Argon Green Laser (Coherent, Palo Alto; CA, USA).
36. 50 μl Hamilton syringe (Hamilton Co.) and a 30-gauge 19° beveled needle (Hamilton Co.).
37. Isopropyl alcohol swabs (Fisher Scientific, Inc.).
38. Dissection equipment: scalpel, jewelers forceps, Castroviejo scissors (Miltex, Inc., York, PA).
39. 37 % formaldehyde solution (Sigma).
40. Bovine serum albumin (BSA) (Sigma).
41. FITC-conjugated isolectin B_4 (L2895, Sigma).
42. Triton X-100 (Sigma).
43. Clear glass slides and Gel/Mount slide coverslipping medium with anti-fading reagents (#M01, Biomedia).

3 Methods

The methods below outline (1) immunomagnetic isolation of EPCs from bone marrow, (2) intracellular loading of EPCs with QD, (3) monitoring of QD loading efficiency, and (4) imaging of QD-loaded EPCs in a rat model of LCNV. While these methods were specifically used for imaging EPCs in LCNV, the techniques herein could be applied toward the labeling and imaging of EPCs in other vascular diseases, including cancer and diabetes, by selection of a relevant animal model.

3.1 Immunomagnetic Isolation of Endothelial Progenitor Cells from Bone Marrow

Cells were isolated from six 4-week-old Brown Norway rats. The animals were sedated with isoflurane (Terrell) vapors, and then euthanized by decapitation. Their tibias and femurs were removed, and the marrow was isolated from these bones in 20 mL total volume of Hank's Buffered Salt Solution (HBSS) [5 mM KCl, 0.44 mM KH_2PO_4, 137 mM NaCl, 0.34 mM Na_2HPO_4, 3.3 mM $NaHCO_3$, 5.5 mM d-glucose, pH 7.1–7.4]. The marrow tissue was then triturated and digested in Accutase™ (A6964, Sigma) for 5 min. Following the digestion, the dissociated cells were washed three times in HBSS followed by centrifugation to remove extracellular matrix debris, and were filtered through a double layer of sterile 40 nm nylon mesh (Sefar America, Inc., Fisher Scientific, Hanover Park, IL). Following this step, the remainder of the procedure was carried out under a laminar flow hood using aseptic technique.

The cell mixture was then purified in order to enrich for $CD34^+$ cells using an immunomagnetic purification method similar to that employed by Su X. et al. for the isolation of $CD31^+$ endothelial cells [16]. We used the EasySep® cell isolation system (18558, StemCell Technologies, *see* **Note 1**). Initially, cells were counted using a hemacytometer to ensure that no more than 2×10^8 cells per milliliter were in an isolation tube (352058, Becton Dickinson). These cells were then centrifuged and resuspended in 1.0 mL of an isolation buffer [3.2 mM Na_2HPO_4, 0.5 mM KH_2PO_4, 1.3 mM KCl, 135 mM NaCl, pH 7.4, 2 % fetal bovine serum (FBS), 1.0 mM EDTA]. An FcR-blocking antibody (112-001-008, Jackson ImmunoResearch Laboratories, Inc.) was then added at 50 μg/ml. Next, 3.0 μg/ml of a rabbit anti-CD34 FITC-conjugated antibody (252268, Abbiotec) was added and the mixture was allowed to incubate for 15 min at room temperature. Following this incubation, the EasySep® FITC Selection Cocktail (18152, StemCell Technologies) was added at 100 μl/ml cells and allowed to incubate for another 15 min at room temperature. Finally, the magnetic nanoparticles (18150, StemCell Technologies) were mixed in at 50 μl/ml cells and allowed to incubate for 10 min. The cell suspension was then brought to a total volume of 2.5 mL with isolation buffer and gently mixed in the tube to ensure a homogenous suspension (*see* **Note 2**). The cells were then placed into the EasySep® magnet (18000, StemCell Technologies) and allowed to sit for 5 min, before the tube contents were carefully discarded into a waste container. This resuspension/magnetic pelleting/discarding process was repeated twice. After the third repetition of this process, the cells were resuspended in EGM-2 media (CC-3162, Clonetics) containing 10 % fetal bovine serum, 1 μg/ml hydrocortisone, 10 ng/ml epidermal growth factor, and 5 ml/500 ml of 100× antibiotic/antimycotic solution (A5955, Sigma) and plated onto 60 mm plastic dishes to be expanded for one full passage before further use.

Prior to in vitro imaging of QD in EPCs, EPCs were seeded at 5,000 cells per well on 4-well Chamber Slides™ (177429, Lab-Tek, *see* **Note 3**) and incubated for 2–4 days at 37 °C until 60 % confluence was achieved. For injection of QD-labeled EPCs in the LCNV model (3.4), EPCs were maintained in 60 mm cell culture dishes until detachment in Accutase and resuspension of EPCs to the desired density.

3.2 Conjugation of acLDL to QD

In order to produce fluorescently labeled acLDL, 100 μl of human low-density lipoprotein (LDL) (5 mg/mL) was added to 10 μl of a 20 mM solution of sulfo-NHS-biotin in borate buffered saline (pH = 8.2, *see* **Note 4**). The mixture was allowed to react for 30 min at room temperature. Following this reaction, the sample was purified for removal of excess reagent using G-25 Sephadex gravity column filtration with borate buffered saline as the elution buffer.

To perform this step, the column buffer is eluted by discarding the column buffer and equilibrating the column with three column volumes of borate buffered saline. These eluents are discarded, and following the third elution step, the biotinylated LDL is pipetted gently on the top of the gel matrix and allowed to penetrate the gel. Next, the column is filled with one column volume of borate buffered saline, and the eluent is collected in 100 μL fractions. acLDL elutes in collections 1–3 due to its larger molecular weight compared to sulfo-NHS-biotin. The remainder of the eluent is discarded. The protein is now purified from excess labeling reagent in preparation for the next step. Following purification, 1.0 mg of sulfo-NHS acetate was added to the product diluted in 1.0 mL total borate buffered saline and the solution was incubated for 1 h. The conjugated protein was then purified using a G-25 column, with borate buffered saline as the elution buffer, to remove excess acetylation reagent, yielding a pure solution of biotinylated acLDL. The conjugated protein was then incubated with 5.0 μl of streptavidin-functionalized QD655 (QD emitting 655 nm light) for 30 min at 37 °C. This solution was characterized for conjugation efficiency by size exclusion chromatography on Sephacryl HR400 resin and was found to exhibit three acLDL molecules per QD.

3.3 Intracellular Loading of EPCs with acLDL-QD

In order to label the EPCs with QD-acLDL and DiI-acLDL (BT-902, Biomedical Technologies, Inc.), the cells were incubated with 2.0 μg/ml of each acLDL in growth media for 6 h at 37 °C. Equivalency between DiI-acLDL and QD-acLDL can be achieved by our knowledge of the 3:1 acLDL:QD ratio. Fifteen minutes prior to ending the incubation, DAPI (Sigma, D9532), a nuclear counterstain, was added to the cells at a concentration of 1 μg/mL. Following the 6 h incubation, the media was removed, and the cells were washed thrice in pre-warmed (37 °C) PBS. After the wash, the cells were coverslipped to be visualized by fluorescence microscopy using Gel/Mount.

3.4 Monitoring of QD-Loading Efficiency

The EPCs were imaged on Chamber Slides™ (177429, Lab-Tek) labeled directly with QD-acLDL and DiI-acLDL using fluorescence microscopy. These cells can be imaged on an upright or inverted fluorescent confocal or wide-field microscope that has the appropriate filter set to distinguish the 568 nm fluorescence of the DiI, the 655 nm fluorescence of the QD, and the 454 nm fluorescence of the DAPI without significant spectral overlap. In our experiments, we used an AX70 microscope (Olympus, Japan) containing filter cubes with 460/40 nm (i.e., corresponding to a 440–480 nm passband of emitted light collected by the camera), 570/30 nm, and 655/20 nm emission filters. Images of the cells were captured using a digital camera attached to the Provis system (DP71, Olympus, Japan) coupled to a computer with image capture software (DP Controller, Olympus, Japan). However, this analysis can be completed with a comparable imaging system.

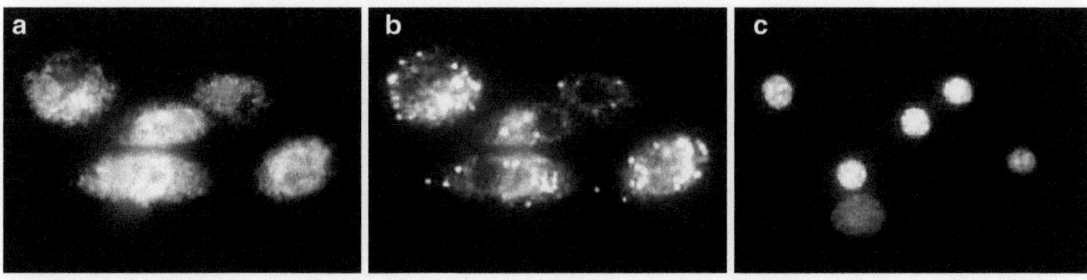

Fig. 1 Imaging of CD34+ EPCs with fluorophore-acLDL conjugates. Panel (**a**) shows the internalized DiI-acLDL conjugate throughout the cytoplasm. Panel (**b**) shows the same field of cells as (**a**), but with internalized, cytoplasmically distributed QD655-acLDL. Panel (**c**) exhibits the DAPI-labeled nuclei of the cells in the same field of view. These panels demonstrate that DiI-acLDL and QD655-acLDL are targeted to and taken up by the same EPC populations. Subsequent analysis by fluorescence microscopy indicated that cells labeled with DiI became dim within 6 min of continuous illumination, but that QD-labeled cells lost only 3–5 % of initial fluorescence for every 10 min of continuous illumination during a 60 min analysis

3.4.1 Imaging of the acLDL-Labeled EPCs

1. Using a microscope configured for excitation with 520/20 nm light, excite the cells labeled by DiI-acLDL and collect an image of resulting fluorescence emission of cells with the 570/30 nm band-pass filter. Capture this image first, since DiI is the most sensitive fluorophore to photobleaching.
2. Without moving the slide, capture the next image of the DAPI-stained cell nuclei using a 360/20 nm excitation light with the 460/40 nm band-pass emission filter.
3. Finally, capture the QD-acLDL-labeled cells using the 360 nm incident light (or if not available, an excitation wavelength of <480 nm) with the 655/20 nm band-pass filter. Note that the smaller wavelength incident light will produce a stronger excitation of the QDs. For optimal results, illuminate the QD-loaded cells for at least 2 min using these incident light settings to induce maximal QD fluorescence emission.

The images shown in Fig. 1 demonstrate that EPCs can be labeled with QD-acLDL with equal efficiency to DiI-acLDL labeling; however, the QD-acLDL labeling results in bright and photostable EPCs that can be tracked in culture for longer periods of time. Specifically, we imaged EPCs labeled with QD continuously up to 60 min with only a minor decrease in intensity, whereas DiI-labeled EPCs after the same illumination period were not visible in the DiI channel.

3.5 Analysis of QD-Loaded EPCs in a Rat Model of Ocular Angiogenesis

3.5.1 Generation of LCNV Model

In this study, CNV, a subretinal form of angiogenesis in which choroidal blood vessels are stimulated to abnormally grow beneath the retinal photoreceptors, was produced in eight Brown Norway rats, each weighing 100 g. This CNV was produced by administering laser radiation between the major retinal vessels of the fundus as we and others have previously published [17–20]. The animals were anesthetized with an IP injection of 80/12 mg/kg ketamine/

xylazine (K113, Sigma), treated with proparacaine (0.5 %) drops for corneal anesthesia, and the pupils dilated with phenylephrine (2.5 %) and atropine sulfate (1 %) drops. A handheld coverslip and (2.5 %) Gonak solution (Akorn) were used as a contact lens for the maintenance of corneal clarity during photocoagulation. Animals were positioned before a slit lamp (Carl Zeiss Meditec, Inc.; Jena, Germany) laser delivery system. An argon green laser (Coherent, Palo Alto; CA, USA) was used for photocoagulation (532 nm wavelength; 360 mW power; 0.07-s duration; and 50 μm spot size). The laser beam was focused on Bruch's membrane with the intention of rupturing it, as evidenced by subretinal bubble formation without intraretinal or choroidal hemorrhage at the lesion site. Each lesion was made in the regions between the major vessels of the retina for a total of six lesions per eye concentrically applied approximately two optic discs from the center. The choroidal capillaries proliferated through the break in Bruch's membrane into the disrupted outer layers of the retina. Laser rupture sites that had subretinal bleeding at the time of lesion creation were excluded from analysis, and represented less than 10 % of total rupture sites in each treatment group. With minor operator training, this procedure can be completed efficiently and reliably such that up to 20 rats per day can be induced for LCNV.

3.5.2 Injection of QD-acLDL-Labeled EPCs into LCNV Model

Seven days after laser treatment, the animals received a tail vein injection of 1×10^6 EPCs labeled with QD655-acLDL in a volume of 40 μl (PBS vehicle) using a 50 μl syringe (Hamilton Co.; Reno, NV) and a 30-gauge needle with a 19° bevel. These cells had been labeled as previously described in Subheading 3.3. For the injection, the rats were anesthetized by isoflurane (Terrell) inhalation, and the tail was wiped with an isopropyl alcohol swab. The injection was made in either of the lateral veins along the base of the animal's tail. As the needle was advanced for the injection, the plunger was pulled back slightly in order to see a flash of blood in the syringe, ensuring the needle was properly in the vein. Fourteen days following laser application, animals were sacrificed by cervical dislocation to prepare ocular tissues for *ex vivo* analysis.

3.5.3 Characterization of EPCs in Retinal Neovascular Tissue

After the animals were sacrificed, eyes were immediately enucleated (removed from the orbit) and placed in fixative, and the cornea, iris, and lens were removed. The retina was peeled away from the fundus gently using a hemostat to reveal the choroid-sclera-RPE section. This section was flattened on a slide, mounted in Gel/Mount, and coverslipped. Endothelial cells and extracellular matrix components were stained to reveal the areas of neovascularization. Areas of abnormal vascular growth were measured via computer-assisted image analysis using high-resolution digital images of the stained sclera-choroid-RPE flat-mounts. The details of this process are described below.

Dissection and Analysis of Neovascularization

1. Enucleate the eyes and place them into labeled containers with 10 % neutral buffered fixative solution [90 % v/v PBS and 10 % v/v 37 % formaldehyde solution] for 2 h at 4 °C.
2. Isolate the choroid, sclera, and RPE. Make sure to remove the entire retina from the RPE surface and remove the cornea, lens, and iris.
3. Place each dissected sample into a labeled 24-well plate well for staining.
4. In the well, wash the samples with PBS for 5 min.
5. Using 5 % BSA in PBS, block the samples on a shaker for 1 h at room temperature.
6. Rinse each sample with PBS three times for 5 min each.
7. Completely cover and submerge each dissected sample with a solution of PBS with 4.0 μg/ml FITC-conjugated isolectin B_4 (L2895, Sigma) and 0.1 % Triton X-100, and incubate for 24 h at 4 °C on shaker.
8. Remove the staining solution and wash with PBS three times for 5 min each.
9. Mount the choroidal flat-mounts on the slides, attempting to make them as flat as possible on the glass. Several cuts from the periphery of the sclera toward the center of the eye will facilitate this process. Use Gel/Mount to prevent the FITC from bleaching (*see* **Note 5**).

Fluorescence Imaging of LCNV Lesions

1. Position slides with flat-mounts into focus using a 20× or 40× objective designed for fluorescence analysis.
2. Using 490/20 nm of incident light, excite the cells labeled by FITC-isolectin B_4 and obscure the fluorescence with the 525/20 nm band-pass filter. Capture this image first, since FITC is the most sensitive fluorophore to photobleaching. Isolectin B_4 labels all endothelial cells within blood vessels, *including* EPCs.
3. Without moving the slide, capture an image of the QD-acLDL-labeled cells using the 360 nm or 490 nm centered excitation light with the 655/20 nm band-pass filter. Note, again, that the smaller wavelength incident light will produce a stronger excitation of the QDs. These settings will reveal the presence of EPCs in the specimen which can be distinguished from other isolectin B_4 positive, mature endothelial cells.

The results of our experiment are shown in Fig. 2 and demonstrate that the QD-acLDL-labeled EPCs make up a large portion (62 % of the total lesion size) of the lesions, and that they frequently coalesce in the middle of the lesion. This observation suggests that

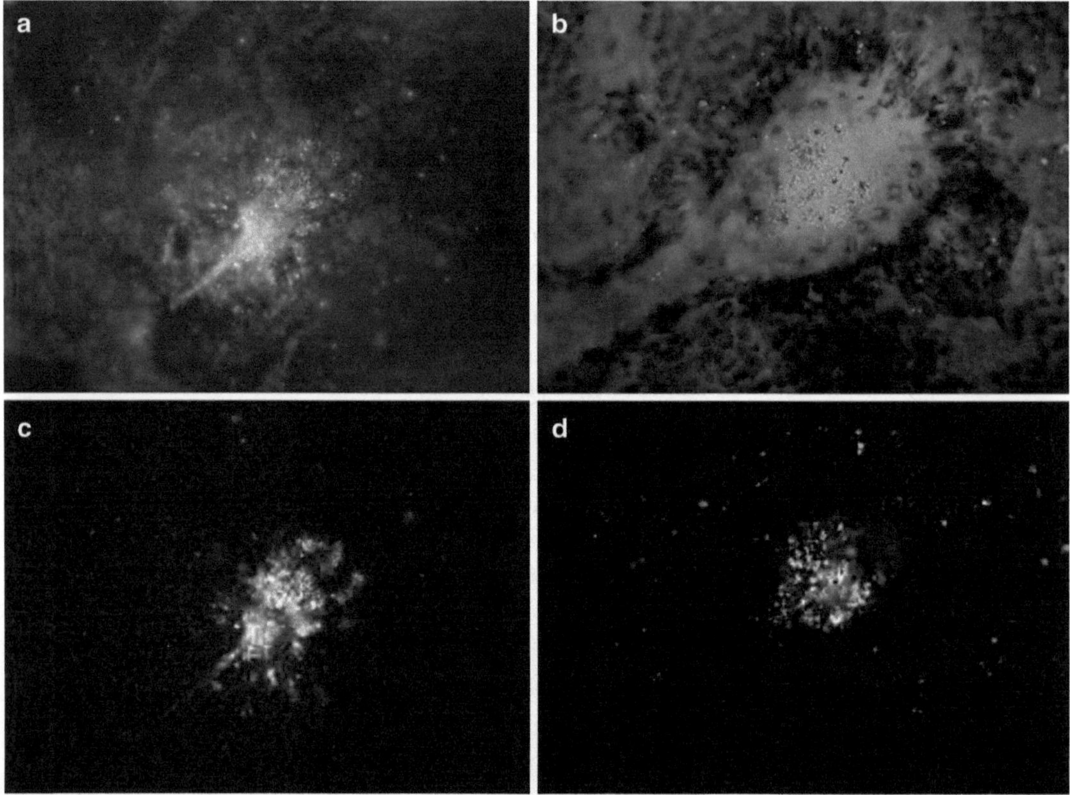

Fig. 2 Imaging of CD34⁺ EPCs labeled with QD655-acLDL in the lesions of LCNV rats undergoing angiogenesis. Panels (**a**) and (**b**) show two different CNV lesions labeled with isolectin B_4-FITC conjugate (all endothelial cells). Panels (**c**) and (**d**) show the QD655-acLDL-labeled $CD34^+$ EPCs from (**a**) and (**b**), respectively, that were injected into the tail vein of LCNV model rats 7 days prior to *ex vivo* analysis. QD-labeled EPCs homed to the neovascular lesions in the choroid in large masses or alternatively arrived in small numbers and proliferated to comprise up to 62 % of neovascular lesion area. This is a clear demonstration that QD-acLDL-labeled EPCs can be reinjected into animals, and tracked for longer periods of time relative to conventional methods, for detailed analysis of EPC function in ocular angiogenesis

EPCs home to the developing lesions through the blood stream and add to the growth of the lesions through their own proliferation as well as the release of growth factors, thereby encouraging neighboring mature endothelial cells to proliferate. This method of cell labeling can be used to efficiently track and study the effects of these EPCs or multiple subtypes of EPCs influencing these neovascular lesions.

4 Notes

1. A number of commercially available kits can be used to enrich $CD34^+$ cell populations. Prominent vendors include Miltenyi, Biotec and Invitrogen. Our experience is limited to the EasySep kit, but other kits may be suitable for this technique, as well.

2. We find that gentle flicking of the tube by hand encourages proper mixing, but vortexing the tube on a low setting is also suitable. Vortexing at medium to high settings is not recommended.
3. The choice of the Lab-Tek chamber slide (Nunc Brand) is intended to allow high-resolution imaging of the labeled EPCs using a fluorescence microscope. The key to successful image acquisition of QD-labeled EPCs is to culture cells on German borosilicate #1.5 glass surfaces treated for cell culture. Other choices include MatTek glass bottom dishes (MatTek, Inc.).
4. Borate buffers have long been established as the most suitable buffers for QD-based fluorescence imaging. Substitution of this buffer in QD conjugation reactions for others, such as PBS, may result in degradation of fluorescence, or aggregation of QD over time.
5. While in our experience Gel/Mount is a suitable mounting medium for imaging QD in tissues, a new mounting media, QMount (Invitrogen), has recently become commercially available. The intended use of this medium is for mounting of biological specimens in a solvent-based solution more appropriate for preservation of QD fluorescence. However, we have not yet observed ocular QD-labeled specimens in QMount at this time.

Acknowledgments

This work was supported by the NIH grants P30EY008126 (Core Grant in Vision Research), AG031036 (JMB), and EY007533 (JSP), a Challenge Award from Research to Prevent Blindness, and grants from the OneSight Foundation (JSP) and the International Retinal Research Foundation (AJ).

References

1. Barnett JM, McCollum GW, Penn JS (2010) Role of cytosolic phospholipase A(2) in retinal neovascularization. Invest Ophthalmol Vis Sci 51:1136–1142
2. Jayagopal A, Russ PK, Haselton FR (2007) Surface engineering of quantum dots for in vivo vascular imaging. Bioconjug Chem 18:1424–1433
3. Sengupta N, Caballero S, Mames RN, Butler JM, Scott EW, Grant MB (2003) The role of adult bone marrow-derived stem cells in choroidal neovascularization. Invest Ophthalmol Vis Sci 44:4908–4913
4. Asahara T, Murohara T, Sullivan A et al (1997) Isolation of putative progenitor endothelial cells for angiogenesis. Science 275:964–967
5. Chan-Ling T, Baxter L, Afzal A et al (2006) Hematopoietic stem cells provide repair functions after laser-induced Bruch's membrane rupture model of choroidal neovascularization. Am J Pathol 168:1031–1044
6. Sengupta N, Caballero S, Mames RN, Timmers AM, Saban D, Grant MB (2005) Preventing stem cell incorporation into choroidal neovascularization by targeting homing and attachment factors. Invest Ophthalmol Vis Sci 46:343–348

7. Peled A, Grabovsky V, Habler L et al (1999) The chemokine SDF-1 stimulates integrin-mediated arrest of CD34(+) cells on vascular endothelium under shear flow. J Clin Invest 104:1199–1211
8. Avigdor A, Goichberg P, Shivtiel S et al (2004) CD44 and hyaluronic acid cooperate with SDF-1 in the trafficking of human CD34+ stem/progenitor cells to bone marrow. Blood 103:2981–2989
9. Zhang ZG, Zhang L, Jiang Q, Chopp M (2002) Bone marrow-derived endothelial progenitor cells participate in cerebral neovascularization after focal cerebral ischemia in the adult mouse. Circ Res 90:284–288
10. Lima e Silva R, Shen J, Hackett SF et al (2007) The SDF-1/CXCR4 ligand/receptor pair is an important contributor to several types of ocular neovascularization. FASEB J 21:3219–3230
11. Yin AH, Miraglia S, Zanjani ED et al (1997) AC133, a novel marker for human hematopoietic stem and progenitor cells. Blood 90:5002–5012
12. de Wynter EA, Buck D, Hart C et al (1998) CD34+AC133+ cells isolated from cord blood are highly enriched in long-term culture-initiating cells, NOD/SCID-repopulating cells and dendritic cell progenitors. Stem Cells 16:387–396
13. Takahashi T, Kalka C, Masuda H et al (1999) Ischemia- and cytokine-induced mobilization of bone marrow-derived endothelial progenitor cells for neovascularization. Nat Med 5:434–438
14. Rafii S (2000) Circulating endothelial precursors: mystery, reality, and promise. J Clin Invest 105:17–19
15. Rafii S, Lyden D (2003) Therapeutic stem and progenitor cell transplantation for organ vascularization and regeneration. Nat Med 9:702–712
16. Su X, Sorenson CM, Sheibani N (2003) Isolation and characterization of murine retinal endothelial cells. Mol Vis 9:171–178
17. Toma HS, Barnett JM, Penn JS, Kim SJ (2010) Improved assessment of laser-induced choroidal neovascularization. Microvasc Res 80:295–302
18. Yanni SE, Barnett JM, Clark ML, Penn JS (2009) The role of PGE2 receptor EP4 in pathologic ocular angiogenesis. Invest Ophthalmol Vis Sci 50:5479–5486
19. Kaplan HJ, Leibole MA, Tezel T, Ferguson TA (1999) Fas ligand (CD95 ligand) controls angiogenesis beneath the retina. Nat Med 5:292–297
20. Bora PS, Hu Z, Tezel TH et al (2003) Immunotherapy for choroidal neovascularization in a laser-induced mouse model simulating exudative (wet) macular degeneration. Proc Natl Acad Sci USA 100:2679–2684

Chapter 5

Imaging Single Synaptic Vesicles in Mammalian Central Synapses with Quantum Dots

Qi Zhang

Abstract

This protocol describes a sensitive and rigorous method to monitor the movement and turnover of single synaptic vesicles in live presynaptic terminals of mammalian central nerve system. This technique makes use of fluorescent semiconductor nanocrystals, quantum dots (Qdots), by their nanometer size, superior photoproperties, and pH-sensitivity. In comparison with other fluorescent probes like styryl dyes and pH-sensitive fluorescent proteins, Qdots offer strict loading ratio, multi-modality detection, single vesicle precision, and most importantly distinctive signals for different modes of vesicle recycling. This application is spectrally compatible with existing optical labels for synapses and thus allows multichannel and simultaneous imaging. With easy modification, this technique can be applied to other types of cells.

Key words Quantum dot, Synaptic vesicle imaging, Fluorescence microscopy, Electron microscopy

1 Introduction

Recent advance in microscope, imaging acquisition and labeling strategy have made it possible to detect, identify and monitor individual proteins and cellular organelles with spatial precision beyond the limit of optical resolution. A wave of ultra-resolution studies using single molecule imaging have brought a whole new view of biological complexity and dynamics previously obscured by ensemble averaging in conventional biochemical approaches [1]. So far, two major types of single molecular imaging techniques have generated the most interests. One is using highly specialized optics or fluorescent probes that allows the detection of all tagged molecules in the field of imaging. Typical examples include STochastic Optical Reconstruction Microscope (STORM) and Stimulated Emission Depletion microscope (STED). The other uses relatively conventional optical setting and mainly relies on mathematical fitting to locate individual fluorescent particles. One representative technique is Fluorescence Imaging with One Nanometer Accuracy

Sandra J. Rosenthal and David W. Wright (eds.), *NanoBiotechnology Protocols*, Methods in Molecular Biology, vol. 1026, DOI 10.1007/978-1-62703-468-5_5, © Springer Science+Business Media New York 2013

(FIONA) introduced by Paul Selvin's group [2]. Although the first one generates extremely detailed view of all labeled molecules regardless their density, it requires special fluorescent probes and offers relatively poor temporal resolution. For higher temporal resolution especially in live-cell imaging, the second approach is more favorable so far. Introduced in 1980s, the precision of single particle tracking has evolved quickly thanks to the development of more sensitive photodetectors and brighter probes, both of which significantly improve the signal-to-noise ratio (S/N). Such improvement is essential since locating single light spots with nanometer precision is not limited by diffraction but rather S/N [2, 3]. For instance, newly invented Electron-Multiplying Charge-Coupled Device (EMCCD) can significantly improve spatial resolution owing to its uncomparable S/N.

One of the most exciting developments in the arena of fluorescent probes is the introduction of fluorescent semiconductor quantum dots (Qdots). Qdots are single particle composed of a crystal core and shell made of semiconductor material such as cadmium selenide and zinc sulfate, respectively. As the core size varies between 2 and 10 nm, their emission increases in parallel, also known as size-tunable emission. For biological applications, Qdots are coated with hydrophilic materials to make them water-soluble and to allow conjugation with biomolecules such as proteins and nucleotides. Qdots offers unique physical and optical features that particularly desired in single-molecular imaging. First, their nanometer size allows Qdot-based probes to access crowded cellular spaces such as the synaptic cleft, which is only a few dozen nanometers at its widest part [4]. Second, they possess photoluminescence hundreds of times brighter than organic fluorophores, which translates to two orders improvement in precision of single particle tracking. Third, Qdots are almost impossible to be photobleached, thereby permits much longer period of image acquisition. Fourth, multicolor imaging is much easier using Qdots because of their broad absorbance and narrow emission. Fifth, Qdots are electron denser than biological structures and therefore offer possibilities of correlated optical and electron microscopy imaging for ultrastructural studies.

Here, we present a standard protocol of Qdot-based single synaptic vesicle imaging that takes advantage of all the aforementioned benefits of Qdot.

So, what is a synaptic vesicle and why it is so important to study single synaptic vesicle behavior? Chemical synapse is specialized cell connections conducting over 95 % of rapid communication between excitable cells. It composes with presynaptic terminal which contains hundreds synaptic vesicles, tiny membrane-bounded organelles filled with neurotransmitter molecules at high concentration [5, 6], and postsynaptic density, which contains receptors for those neurotransmitters. Whether the communication is neuron-to-neuron or neuron-to-muscle, or excitatory, inhibitory, or

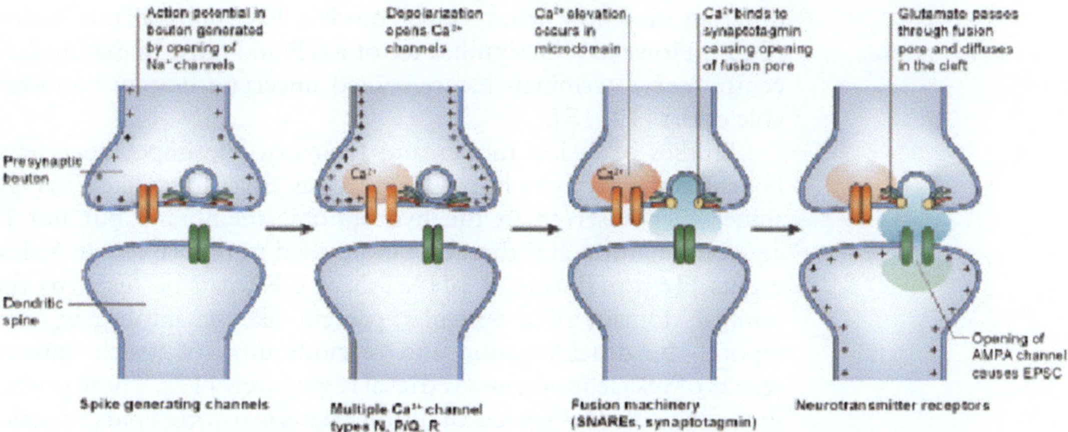

Fig. 1 Steps in the process of chemical synaptic transmission, showing basic agreement regarding millisecond events in neurotransmission and their molecular basis. Similar events occur at various synapses in both vertebrates and invertebrates but are illustrated for the case of a glutamatergic synapse of the mammalian CNS. From ref. [7]

modulatory, fundamental aspects of the unitary signaling event appear to be similar and highly refined. The sequence of events that underlie such quantal transmission has been studied heavily in central excitatory synapses, typical of those that support sensation, action, learning and memory in the brain (Fig. 1) [7]. The early steps include the arrival of membrane depolarization at the presynaptic structure (nerve terminal or synaptic bouton) and the opening of voltage-gated Ca^{2+} channels. Ca^{2+} influx through these channels leads to a surge of Ca^{2+} in nanodomains near the presynaptic release machinery [8], which includes SNARE proteins and the Ca^{2+} sensors like Synaptotagmin I [9]. Conformational change of the Ca^{2+} sensors causes the fusion between vesicular membrane and plasma membrane, so called exocytosis, which opens vesicular lumen to the extracellular space. The diffusional spread of neurotransmitter molecules through such opening finally results in rapid activation of ionotropic receptors residing on postsynaptic cell membrane, which convert chemical signal (neurotransmitter) back to the change of membrane potential. Clearly, the central event linking the early and late steps is neurotransmitter release, wherein synaptic vesicles discharge their contents into the narrow cleft between the presynaptic and postsynaptic membranes. This requires the establishment of continuity between the vesicle interior and the cleft via a passageway called the fusion pore [10, 11]. The nature of the continuity has important implications for how the contents of the vesicle are released and for the fate of the vesicle membrane, topics of intensive research that is still ongoing. Among various models on vesicle fusion, an unconventional one known as kiss-and-run (K&R), the vesicle is thought to release its contents, yet retain enough identity to be reused again, may carry different function from other models like full-collapse fusion (FCF). However, the significance of K&R and reuse in mammalian central nerve terminals has remained uncertain despite considerable effort [12–15].

Studies of fusion modes and their possible importance have benefited greatly from fluorescent probes. Styryl dyes like FM1-43 shine when inserted in the hydrophobic membrane but not in aqueous solution and thereby can be used to report vesicle fusion events [16]. pHluorin, a pH-sensitive GFP, can be fused to the luminal domain of a vesicular protein like synaptobrevin, and reports the deacidification and reacidification of vesicle lumen, results of vesicle fusion and retrieval respectively [17]. These probes focus on particular aspects of fusion events involving lipids or vesicular proteins, and have supplied valuable insights into vesicle dynamics. However, neither FM dyes nor pHluorin fusion proteins provide a signal that can directly distinguish K&R from FCF. FM dye can leave the vesicle membrane by various routes, by aqueous departitioning and escape through the fusion pore during K&R [14, 18], or by lateral diffusion away from fused membrane upon FCF [19]. Clusters of pHluorin fused vesicular proteins can persist in two ways, corralled in the vesicle during K&R, or tethered by auxiliary proteins even after FCF [20]. More importantly, these probes do not reporting the change of vesicle shape, fundamental to the original definitions of FCF and K&R [21–23].

To describe the dynamic properties and functional impact of K&R, the field desires a more direct and reliable method. We sought an approach to discriminate sharply between FCF and K&R, focusing on this fundamental distinction: the degree of opening of the vesicle lumen. Inspired by the finding that chromaffin granules use K&R to release small molecules (catecholamines) and FCF to discharge their dense peptide core [24], we used an artificial cargo of appropriate size: quantum dots (Qdots). Like probes for sizing the pore regions of ion channels, Qdots could gauge the narrowest aperture in the path between vesicle lumen and external medium, escaping only if the vesicle undergoes drastic loss of shape, i.e., FCF.

Other than their superior photoproperties, the emission intensity of Qdots is sensitive to environmental factors such as pH [25]. Qdots emitting at 605 nm have a narrow emission spectrum that fits neatly between the emission peaks of EGFP and FM4-64. This allows concurrent visualization of Qdots and the other optical probes. With an organic coating bearing carboxyl groups, the Qdots have a hydrodynamic diameter of 15 nm, determined with quasi-elastic light scattering [26, 27]. This size is appropriate in multiple respects (Fig. 2): small enough for only one Qdot to fit into a synaptic vesicle, but large enough to be completely rejected by K&R fusion pores (1–3 nm) [28]. According to these unique features, single Qdots residing in individual vesicles provided two sharply

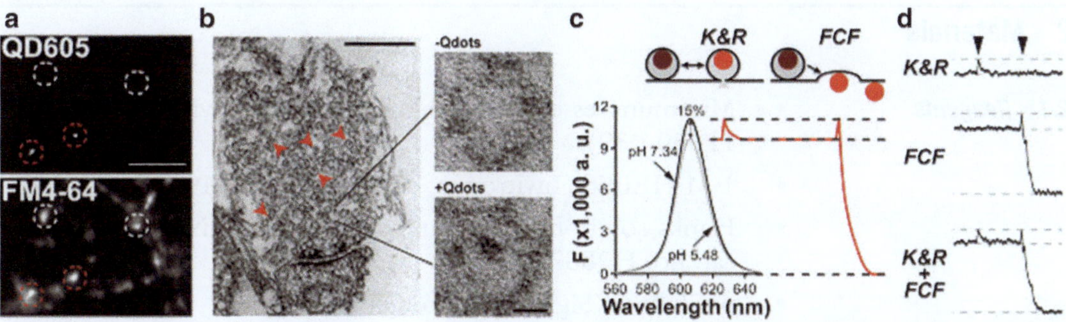

Fig. 2 (**a**) Qdot-labeled single vesicles in presynaptic terminals outlined by subsequent FM4-64 staining. Scale bar, 3 μm. (**b**) Electron micrographs show Qdots loaded into synaptic vesicles. Scale bars, 400 nm (*right*) and 10 nm (*left*). (**c**) Spectrophotometer measured Qdot fluorescence change at different pHs and hypothesized Qdot signals corresponding to K&R and FCF. (**d**) Sample recordings of field stimulus (*arrowhead*) evoked changes of Qdot fluorescence

distinct signals for K&R and FCF (Fig. 2), enabling us to show how the prevalence of K&R and the duration of fusion pore opening are regulated by firing frequency, and how vesicles are re-acidified before undergoing reuse [27]. Furthermore, individually Qdot-labeled synaptic vesicles can be followed with nanometer precision and thus yield valuable information about vesicle mobility within and between adjacent synapses prior to and immediately after fusion. Last but not the least, correlated investigation using electron microscopy can reveal the geometric relationship between recycling vesicles and active zones at presynaptic terminals (i.e., labeled with Qdots), an essential indicator for the kinetics of neurotransmitter release.

The current protocol has two major limitations. First, the mechanism of vesicular uptake of Qdots, especially what determines the Qdots affinity to neuronal membrane is not complete understood. We know that either a negative charge or carboxyl group itself is necessary for such affinity. But we do not know the target of the attachment and thus lack an effective way to control this membrane affinity. However, by systematic adjustment of coating properties, it is possible to optimize this key parameter such that Qdots can bind to neuronal membrane but not too tight such that they can readily departure from the membrane after exiting vesicles. The second limitation results from the intrinsic photo-intermittency of Qdots, a.k.a. "blinking." Although blinking events provides an independent criterion to quantify unitary Qdot fluorescence, the randomness of blinking duration and intervals obscure fluorescence measurement. To solving this problem, specific algorithms for overcoming blinking events are rapidly evolving [29]. With the continuous development of Qdots and the advancement of super-resolution imaging techniques, it is certain that Qdot-based single molecule imaging will become a breakthrough in cellular Neuroscience.

2 Materials

2.1 Reagents

- Minimum essential medium (MEM) (Invitrogen, cat. no. 11360-039).
- 1 M HEPES (Invitrogen, cat. no. 15630-049).
- Hanks solution (made directly from pre-mix powder, Sigma, cat. no. H2837).
- D-Glucose (Sigma, cat. no. G7021).
- 200 mM L-glutamine (Invitrogen, cat. no. 25030-032).
- Sodium pyruvate (Invitrogen, cat. no. 11360-039).
- B27 supplement (Invitrogen, cat. no. 17504-044).
- Transferrin (Calbiochem 616424).
- Insulin Sigma I5500.
- 8 μM carboxyl Qdot 605 (Invitrogen, cat. no. Q21301MP).
- 1 M HCl.
- 1 M NaOH.
- $NaHCO_3$.
- NaCl.
- KCl.
- 1 M $MgCl_2$.
- 1 M $CaCl_2$.
- Bovine serum albumin (Sigma, cat. no. A7030).
- Sodium azide.
- 1.43 M sucrose (store at 4 °C).
- Styryl dye FM 4-64: *N*-(3-triethylammoniumpropyl)-4-(6-(4-(diethylamino)phenyl)hexatrienyl)pyridinium dibromide (Invitrogen, cat. no. T3166).

2.2 Equipment

- CO_2 incubator for cell culture.
- Vacuum pump for perfusion.
- Cultured cells on size 0 circular glass coverslips.
- Recording chamber that allows optical imaging and electrical field stimulation (e.g., D6RG chamber and PH1 platform, Wanner Instrument).
- Imaging software (e.g., Metamorph).
- Analytic software (e.g., ImageJ).

2.3 Reagent Setup

- Tyrode's solution (containing, in mM, 150 NaCl, 4 KCl, 2 $CaCl_2$, 2 $MgCl_2$, 10 glucose, 10 HEPES (310–315 mosm), with pH set at 5.48 or 7.35 with NaOH). 5 μM NBQX and 50 μM D-APV are added to prevent recurrent activity and the development of synaptic plasticity during stimulation.

- 90K Tyrode's solution (containing, in mM, 64 NaCl, 90 KCl, 2 $CaCl_2$, 2 $MgCl_2$, 10 glucose, 10 HEPES (310–315 mosm), with pH set at 5.48 or 7.35 with NaOH). 5 μM NBQX and 50 μM D-APV are added to prevent recurrent activity and the development of synaptic plasticity during stimulation.
- Culture medium (in 500 ml MEM, 2.5 g glucose, 100 mg $NaHCO_3$, 50 mg transferrin, 50 ml FBS, 5 ml 0.2 M L-glutamine, 12.5 mg insulin; 0.2 μm filter-sterilize and aliquot in 50 ml tubes).

2.4 Equipment Setup

- Inverted microscope equipped with a high numerical aperture (NA) objective (NA>1.3), an arc lamp or laser and appropriate filter sets. We use inverted microscopes from Nikon (Ti-E) and an oil-immersion objective (Nikon X60 NA=1.49). Excitation and emission filters are as follows: D470/40 excitation and HT605/20 emission for Qdot, D490/20 excitation and D660/50 emission for FM 4-64, and standard GFP filter set for EGFP and pHluorin (all filters are from Chroma Technology Corp.).
- Field stimulation is delivered via a pair of electrodes placed above the imaging stage and controlled by a micro-manipulator. The electrodes are connected to a stimulation isolator (Wanner Instruments). The input channels is connected to a Digidata 200B and controlled by P-clamp software (Molecular Device).

*Critical QDs can be excited at any wavelength. Use a wide band-pass excitation filter to collect maximum light. Avoid excitation with ultraviolet (UV) light, which leads to photodamage of the cells and increased blinking of QDs. A narrow-band emission filter avoids crosstalk of signals from different fluorophore. Sensitive image acquisition system, such as a EMCCD (e.g., iXon+897 from Andor).

3 Procedure

Rat Hippocampal culture (as described in ref. [30] with little modification).

1. Dissect hippocampi from postnatal day 0–2 Sprague-Dawley rats and isolate CA1 and CA3 region. Cut into small pieces (1×1 mm).
2. Wash tissue 3× with Hanks solution and digested with Trypsin (0.025 %).
3. Wash tissue 3× with Hanks solution and titrate with fire polished glass pipette.

4. Centrifuge for 10 min, 800 × *g*, 4 °C.
5. Aspire supernatant and resuspend cell pellet with culture medium.
6. Plate cell on glass coverslips in culture plates or dishes.
7. Add additional culture medium up to 2 ml after 1 h.
8. Add mitosis inhibitor 1–2 day after cell plating and cell will be ready after 12–18 days in vitro.

Purification of Qdot solution

1. Centrifuge with maximum speed for 15 min at 4 °C. *This step moves most of the Qdot aggregates.
2. Gel filtration of Qdot with size exclusion column (e.g., Superdex 200) and collect all eluted solutions.
3. Concentrate it with Vivispin 2 ml by adding all eluted solutions in the concentrator chamber and spin at 10,000 × *g*, 4 °C for 10 min.
4. Remove collection tube and re-suspend Qdots with appropriate volume of Tyrode's to have the final stock solution of Qdot reconstituted at 8 μM.

Labeling synaptic vesicles with Qdots

1. Use bent-tip forceps to carefully pick up coverslips and place it in sealed imaging chamber containing 400 μl Tyrode's solution.
2. Place imaging chamber on the microscope stage and connect the gravity-vacuum perfusion tubing.
3. Turn on perfusion and set dripping speed at 1–2 drops/s.
4. Open transmitter light and search for field of imaging. The criterion is to avoid neuronal soma and area clustered with dendrites and axons.
5. Stop perfusion and keep coverslips submerged in ~400 μl Tyrode's solution.
6. Dilute Qdot stock solution to 80 nM in 400 μl Tyrode's solution for whole bouton labeling or 0.2 nM in 400 μl Tyrode's solution for single vesicle loading.
7. Slowly drop the 400 μl Qdot-containing Tyrode's solution in the imaging chamber and mix gently.
8. Submerge the stimulation electrodes into Tyrode's solution.
9. Deliver a train of field stimulation (e.g., 60-s resting period, 10-Hz 2-min spike train, and 60-s resting period).
10. Switch on perfusion and wash the cells for 15–20 min. To ensure complete removal of unloaded Qdots, periodically examine the Qdot signals using eyepiece. Presence of Qdots outside of dendrite or axon areas is the sign that the wash-off is not complete.

Imaging

1. During the washing period, adjust imaging settings includes exposure time (e.g., 100 ms), frame interval (e.g., 333 ms for 3 Hz imaging rate) and detector gain (i.e., electron-multiplication) such that the maximum fluorescence signal of Qdots reaches 80 % of the dynamic range of the detection device (e.g., EMCCD).
2. Set up the synchronization of imaging rate and stimulation trigger. For example, if field electrical stimulation is applied, set the trigger of field stimulation protocol as the opening of the imaging shutter. In our experiments, we derived a TTL pulse from imaging program upon the first opening of the excitation shutter. This pulse behaves as an external trigger for the start of a 0.1-Hz 2-min field stimulation train with 30-s pre-stimulation delay for baseline measurement.
3. Choose regions of interests (ROIs) for fluorescence measurement. This can be done online or off-line. If the imaging programs allow (e.g., MetaFluor), it is better to do it online so imaging and the first step of data analysis can be conducted simultaneously. We use circular ROIs with fixed positions and a same size cover all synaptic boutons identified by overexpressed vesicular proteins fused with fluorescent proteins or co-labeled with FM4-64. The size of ROIs are ~2 μm, which is twice of the average size of CNS presynaptic terminals. In images, the size of ROIs are measured as the number of pixels. The size of pixels is determined by magnification and chip size of the photon detector.
4. Start continuous imaging and store the imaging stack on the computer RAM. Upon finishing, immediately rename the imaging file encoding experiment information and save it on the hard drive. If online ROI measurement is conducted, export the fluorescence intensity data to appropriate file format (e.g., Microsoft Excel) and save it accordingly.

Data Analysis

1. Single synaptic vesicle tracking. The trajectories of individual Qdot-labeled synaptic vesicles are reconstructed from the image stacks. Single vesicles are identified by its Qdot fluorescence. A variety of open-source software packages are available for such analysis. For example, ImageJ (http://rsbweb.nih.gov/ij/) and ImageJ-based Fiji (http://pacific.mpi-cbg.de/wiki/index.php/Fiji). A Qdot-specialized tracking program, SINEMA (http://www.lkb.ens.fr/recherche/optetbio/tools/sinema/), is used in some of our studies. In brief, this is a two-step process that applied successively to each frame of the image stack. First, fluorescent spots are detected

by cross-correlating the image with a Gaussian model of the Point Spread Function. A least-squares Gaussian fit is employed to determine the center of each spot and the spatial accuracy is determined by signal-to-noise ratio. Second, Qdot trajectories are assembled automatically by linking, from frame to frame, the computer-determined centers from the same Qdots. If blinking occurs, the association criterion is based on the assumption of free Brownian diffusion and missing frames are linked by straight line for simplification. In addition, a post-analysis manual association step is performed to exclude wrongly associated trajectories from neighboring Qdots.

2. Distinguishing single FCF and K&R events. This is a three-step procedure including digitizing fluorescence signal, identifying stimulation-evoked fusion events and spontaneous events, distinguishing FCF and K&R based on the pattern of fluorescence change. First, average fluorescence intensity in the ROIs are extrapolated and stored as absolute numbers in data file. Background fluorescence is measured and averaged from four non-Qdot areas of each frame. The Qdot fluorescence change thus can be extracted by subtracting absolute values with background fluorescence. The pre-stimulation and post-stimulation baseline can be determined by averaging Qdot fluorescence before and after field stimulation. Second, align Qdot signal with field stimulation pulses along timeline. All events that exhibit 15 % Qdot fluorescence jump within 300 ms (i.e., one imaging frame with 3-Hz imaging rate) after field stimulation pulse are designated as evoked fusion events and all other events are determined as spontaneous events. Third, manually identify FCF and K&R events, i.e., transient increase of 15 % unitary Qdot fluorescence as K&R and 15 % transient increase followed by unitary loss of Qdot fluorescence as FCF. All fusion events of every Qdot-labeled can be further digitized for raster plot or further analysis.

4 Anticipated Results

Figure 2 shows an example of cultured hippocampal neurons labeled with Qdots. Presynaptic boutons are indentified with retrospective staining of FM4-64. The signal of FM4-64 and Qdots can be separated using appropriate filter sets. The emission of Qdots and other fluorescence dyes or proteins can also be detected simultaneously using optical devices like Dual-View (Photometrics).

Figure 3 shows sample traces representing single K&R and FCF evoked by field stimulation. In addition, hundreds of events from Qdot-labeled single synaptic vesicles are summarized in digitized

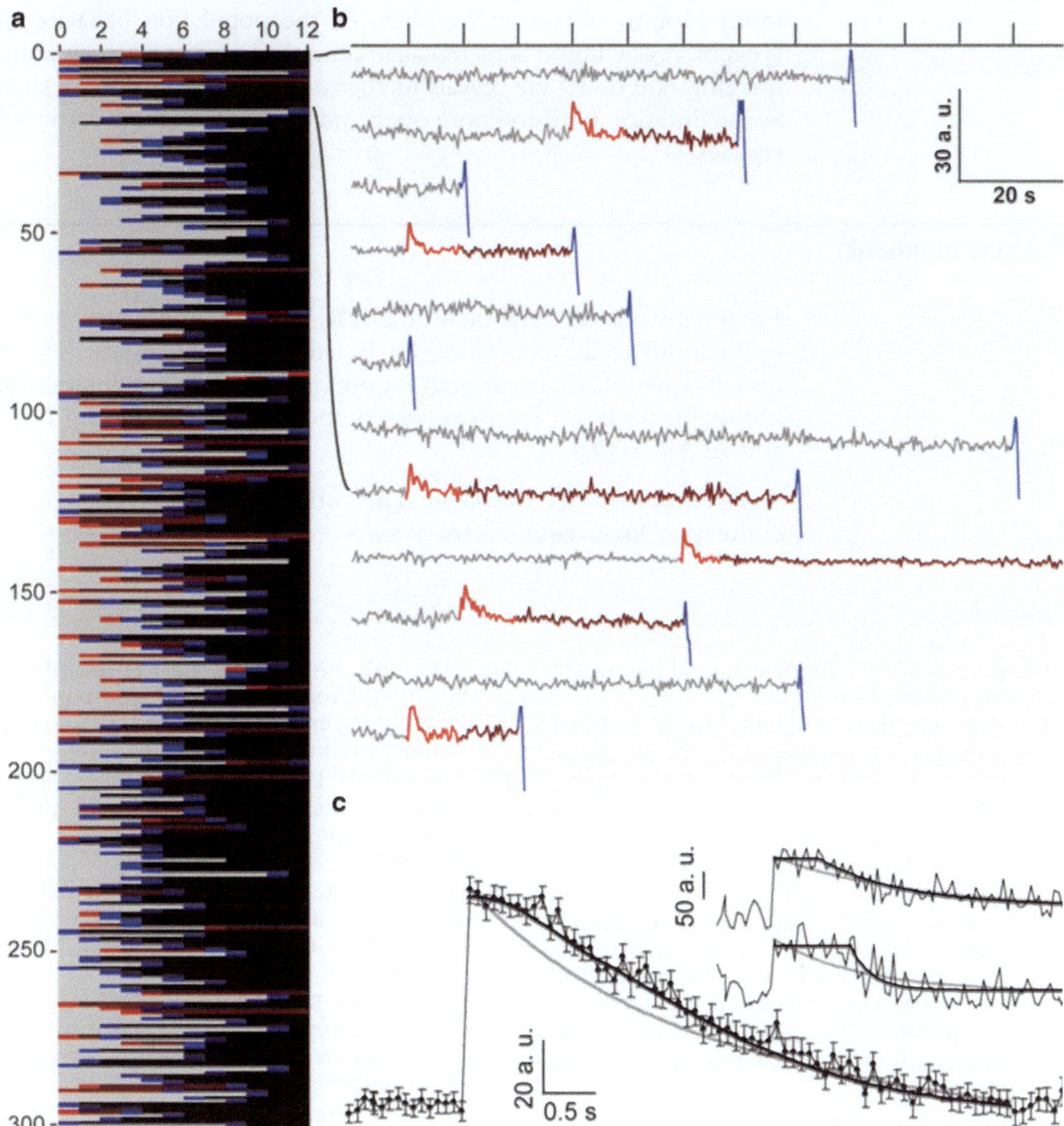

Fig. 3 (**a**) Raster representation of traces ($n=302$) from single Qdot-loaded vesicles that responded to 0.1-Hz field stimulation for 2 min. For each stimulus and subsequent interval, Qdot signals registered as nonresponse (*gray*), K&R (*red*), nonresponse after K&R (*maroon*), FCF (*blue*), or Qdot no longer present in region of interest (*black*). Pooled traces from $N=8$ cover slips, three separate cultures. (**b**) Traces corresponding to the first 12 rasters in (**a**). Photoluminescence changes are color-coded for each stimulus and ensuing interval as in (**a**). (**c**) Comparison of pooled data exemplified by *insets* (*black symbols* with *error bars* indicating SEM, $n=43$) and averages of the two kinds of fits (*gray* and *black*). *Insets*: Samples taken in normal Tyrode's solution with single shocks; interstimulus interval, >20 s. Two types of fits were overlaid: a single-exponential decay (*gray*) and a plateau followed by an exponential (*black*), the latter fitting significantly better even after statistical penalization for the extra parameter (Akaike information criterion score, –60.5; $P<0.001$)

raster plot to obtain an overview of the population behavior of synaptic vesicle fusion at mammalian central nerve system. Increasing imaging rate to 30-Hz reveals further detail of fusion kinetics such as the duration of fusion pore open and re-acidification of retrieved vesicles.

Acknowledgments

I thank my postdoctoral mentor, Dr. R. W. Tsien, for giving me the opportunity to develop this Qdot-based vesicle labeling method in his lab. I also thank Invitrogen for providing the documentation of Qdots properties. This work was supported by grants from NIH and AFAR to Q.Z.

Competing Interests Statement: The author declares that he has no competing financial interests.

References

1. Hell SW (2007) Far-field optical nanoscopy. Science 316:1153–1158
2. Selvin PR, Ha T (2008) Single-molecule techniques: a laboratory manual. Cold Spring Harbor Laboratory Press, Cold Spring Harbor, NY
3. Huang B, Bates M, Zhuang X (2009) Super-resolution fluorescence microscopy. Annu Rev Biochem 78:993–1016
4. Dahan M, Levi S, Luccardini C, Rostaing P, Riveau B, Triller A (2003) Diffusion dynamics of glycine receptors revealed by single-quantum dot tracking. Science 302:442–445
5. Katz B (1969) The release of neural transmitter substances. Charles C Thomas, Springfield, IL
6. Suszkiw JB, Zimmermann H, Whittaker VP (1978) Vesicular storage and release of acetylcholine in Torpedo electroplaque synapses. J Neurochem 30:1269–1280
7. Lisman JE, Raghavachari S, Tsien RW (2007) The sequence of events that underlie quantal transmission at central glutamatergic synapses. Nat Rev Neurosci 8:597–609
8. Llinas R, Sugimori M, Silver RB (1992) Microdomains of high calcium concentration in a presynaptic terminal. Science 256: 677–679
9. Sudhof TC (2004) The synaptic vesicle cycle. Annu Rev Neurosci 27:509–547
10. Jackson MB, Chapman ER (2006) Fusion pores and fusion machines in Ca^{2+}-triggered exocytosis. Annu Rev Biophys Biomol Struct 35:135–160
11. Spruce AE, Breckenridge LJ, Lee AK, Almers W (1990) Properties of the fusion pore that forms during exocytosis of a mast cell secretory vesicle. Neuron 4:643–654
12. Aravanis AM, Pyle JL, Tsien RW (2003) Single synaptic vesicles fusing transiently and successively without loss of identity. Nature 423: 643–647
13. He L, Wu XS, Mohan R, Wu LG (2006) Two modes of fusion pore opening revealed by cell-attached recordings at a synapse. Nature 444:102–105
14. Richards DA, Bai J, Chapman ER (2005) Two modes of exocytosis at hippocampal synapses revealed by rate of FM1-43 efflux from individual vesicles. J Cell Biol 168:929–939
15. Serulle Y, Sugimori M, Llinas RR (2007) Imaging synaptosomal calcium concentration microdomains and vesicle fusion by using total internal reflection fluorescent microscopy. Proc Natl Acad Sci U S A 104:1697–1702
16. Betz WJ, Mao F, Bewick GS (1992) Activity-dependent fluorescent staining and destaining of living vertebrate motor nerve terminals. J Neurosci 12:363–375
17. Miesenböck G, De Angelis DA, Rothman JE (1998) Visualizing secretion and synaptic transmission with pH-sensitive green fluorescent proteins. Nature 394:192–195
18. Aravanis AM, Pyle JL, Harata NC, Tsien RW (2003) Imaging single synaptic vesicles undergoing repeated fusion events: kissing, running, and kissing again. Neuropharmacology 45: 797–813

19. Zenisek D, Steyer JA, Feldman ME, Almers W (2002) A membrane marker leaves synaptic vesicles in milliseconds after exocytosis in retinal bipolar cells. Neuron 35:1085–1097
20. Willig KI, Rizzoli SO, Westphal V, Jahn R, Hell SW (2006) STED microscopy reveals that synaptotagmin remains clustered after synaptic vesicle exocytosis. Nature 440:935–939
21. Ceccarelli B, Hurlbut WP, Mauro A (1973) Turnover of transmitter and synaptic vesicles at the frog neuromuscular junction. J Cell Biol 57:499–524
22. Fesce R, Grohovaz F, Valtorta F, Meldolesi J (1994) Neurotransmitter release: fusion or 'kiss-and-run'? Trends Cell Biol 4:1–4
23. Heuser JE, Reese TS (1973) Evidence for recycling of synaptic vesicle membrane during transmitter release at the frog neuromuscular junction. J Cell Biol 57:315–344
24. Fulop T, Radabaugh S, Smith C (2005) Activity-dependent differential transmitter release in mouse adrenal chromaffin cells. J Neurosci 25: 7324–7332
25. Gao X, Chan WC, Nie S (2002) Quantum-dot nanocrystals for ultrasensitive biological labeling and multicolor optical encoding. J Biomed Opt 7:532–537
26. Zhang Q, Cao YQ, Tsien RW (2007) Quantum dots provide an optical signal specific to full collapse fusion of synaptic vesicles. Proc Natl Acad Sci USA 104:17843–17848
27. Zhang Q, Li Y, Tsien RW (2009) The dynamic control of kiss-and-run and vesicular reuse probed with single nanoparticles. Science 323:1448–1453
28. Jackson MB, Chapman ER (2008) The fusion pores of Ca^{2+}-triggered exocytosis. Nat Struct Mol Biol 15:684–689
29. Bonneau S, Dahan M, Cohen LD (2005) Single quantum dot tracking based on perceptual grouping using minimal paths in a spatiotemporal volume. IEEE Trans Image Process 14:1384–1395
30. Liu G, Tsien RW (1995) Synaptic transmission at single visualized hippocampal boutons. Neuropharmacology 34:1407–1421

Chapter 6

Quantum Dot-Based Single-Molecule Microscopy for the Study of Protein Dynamics

Jerry C. Chang and Sandra J. Rosenthal

Abstract

Real-time microscopic visualization of single molecules in living cells provides a molecular perspective of cellular dynamics, which is difficult to be observed by conventional ensemble techniques. Among various classes of fluorescent tags used in single-molecule tracking, quantum dots are particularly useful due to their unique photophysical properties. This chapter provides an overview of single quantum dot tracking for protein dynamic studies. First, we review the fundamental diffraction limit of conventional optical systems and recent developments in single-molecule detection beyond the diffraction barrier. Second, we describe methods to prepare water-soluble quantum dots for biological labeling and single-molecule microscopy experimental design. Third, we provide detailed methods to perform quantum dot-based single-molecule microscopy. This technical section covers three protocols including (1) imaging system calibration using spin-coated single quantum dots, (2) single quantum dot labeling in living cells, and (3) tracking algorithms for single-molecule analysis.

Key words Quantum dot, Biological labeling, Protein trafficking, Single-molecule, Fluorescence microscopy

1 Introduction

In recent years microscopy techniques in cell biology have seen remarkable progress. Various new methods, such as near-field scanning optical microscopy (NSOM) [1], single-molecule high-resolution imaging with photobleaching (SHRIMP) [2], stimulated emission depletion (STED) microscopy [3], and stochastic optical reconstruction microscopy (STORM) [4], were developed to achieve a significantly higher spatial resolution, which permits discovery of subcellular structures in great detail. Despite the fact that these imaging methods are designed to break the optical diffraction barrier, they are not very accessible to cell biology researchers. Also, the data analyses are usually extremely complicated, requiring well-trained experts in the field.

Sandra J. Rosenthal and David W. Wright (eds.), *NanoBiotechnology Protocols*, Methods in Molecular Biology, vol. 1026, DOI 10.1007/978-1-62703-468-5_6, © Springer Science+Business Media New York 2013

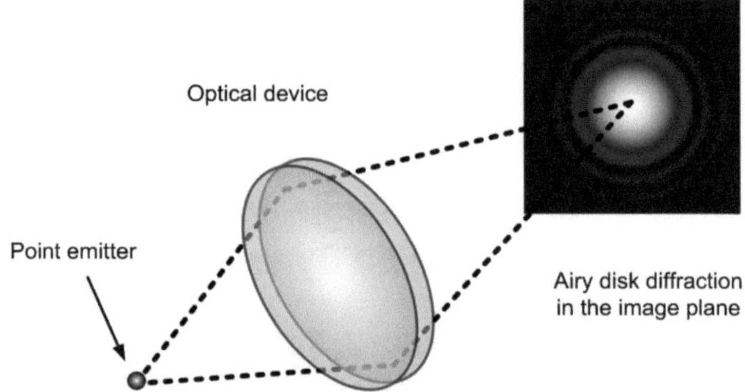

Fig. 1 Schematic representation of the diffraction pattern from a point emitter passed through an optical device. Owing to diffraction, the smallest distance to which an imaging system can optically resolve separate light sources at is limited by the size of the Airy disk

Single-molecule fluorescent microscopy, derived from high-speed fluorescence microscopy, is probably the most accessible method for cell biologists to investigate single-molecule dynamics in living cells. The basis of this method is to follow the real-time movement of individual molecules by using fluorescent microscope, resulting in a map of the dynamics upon observing many individual events. Based on the above definition, an ultimate sensitivity is required, which allows *individual single molecules* to be monitored in a picoliter- to femtoliter-sized microscope sampling volume. However, even if performed on an optimal designed microscope equipped with a camera offering single-photon sensitivity per frame, optical detection of a single molecule is still diffraction limited. The diffraction pattern of a point object viewed through a microscope, known as the Airy disk (Fig. 1), can be modeled by an appropriate point-spread function (PSF). The theoretical 2D paraxial PSF of the wide-field fluorescence microscope can be calculated as [5]

$$\mathrm{PSF} = \left[2 \times \frac{J_1(k_{\mathrm{em}} \times \mathrm{NA} \times r)}{k_{\mathrm{em}} \times \mathrm{NA} \times r} \right]^2, \tag{1}$$

where r is the radial distance to the optical axis, NA is the numerical aperture, J_1 is the first-order Bessel function, $k_{\mathrm{em}} = 2\pi / \lambda$ defines the emission wave number, and λ is the wavelength of light. The smallest resolvable distance between two points of the 2D plane corresponds to the first root of the PSF and is given by the Rayleigh distance [6],

$$d_{xy}^{\mathrm{R}} = 0.61\frac{\lambda}{\mathrm{NA}} \tag{2}$$

As can be seen from Eqs. 1 and 2, under the important prerequisite in which single emitters are placed in extremely low concentration to be spatially separated greater than the diffraction-limited region, it is possible to identify the localization (x, y) of a single molecule from an optical microscope image by fitting the PSF. Indeed, the fundamental idea of the modern single-molecule microscopy is reliant on PSF fitting to localize the centroid position of single-point emitters. Typically, fluorescent intensity distribution from a single emitter can be fit with a 2D Gaussian [7, 8]. To calculate a subpixel estimate of the position from a single emitter, the general method is to fit the intensity distribution with a 2D Gaussian function and then calculate the local maximum intensity:

$$I_{xy} = A_0 + A \times \mathrm{e}^{-\frac{(x-x_0)^2+(y-y_0)^2}{w^2}}, \tag{3}$$

where I_{xy} is the intensity of the pixel, x_0 and y_0 are the designated local maximum coordinates of the Gaussian, A is the amplitude of the signal with local background A_0, and w is the width of the Gaussian curve. The smallest distance at which two emitters can be recognized and separated is roughly equal to the full width at half maximum (FWHM) of the w:

$$w_{\mathrm{FWHM}} = w\sqrt{\ln(4)} = 1.177w \tag{4}$$

The coordinate (x_0, y_0) acquired by fitting function (Eq. 3) using chi-squared minimization is not a true position, but only an estimate. The accuracy is strongly dependent upon the respective signal-to-noise ratio (SNR), which is defined as [8]

$$\mathrm{SNR} = \frac{I_0}{\sqrt{\sigma_{\mathrm{bg}}^2 + \sigma_{I_0}^2}}, \tag{5}$$

where I_0 is the maximum signal intensity above background, σ_{bg}^2 is the variance of the background intensity values, and $\sigma_{I_0}^2$ is the true variance of the maximum signal intensity above the background. Since w width is approximately equal to the wide-field diffraction limit (for visible light is about 250 nm), the uncertainty of the fitted coordinate ($\Delta\sigma$) is approximately given by

$$\Delta\sigma \approx \frac{250}{\mathrm{SNR}}(\mathrm{nm}) \tag{6}$$

However, common organic fluorophores used in single-molecule imaging have very limited photon yield which makes them difficult to produce a sufficient SNR for single-molecule imaging. It is also important to note that organic fluorescent dyes also suffer from narrow Stokes shift (difference in excitation and emission wavelength). This drawback can increase background signals and further reduce SNR. In addition, photobleaching associated with organic fluorophores prevents long-term monitoring in single-molecule microscopy. Hence, the shortcomings associated with organic fluorophores place a high demand for new fluorescent materials for single-molecule tracking.

Over the past decade, quantum dots (qdots) have shown tremendous potential for in vitro and in vivo biological imaging [9–14]. Among the fluorophores developed for single-molecule tracking, a fluorescent qdot is, perhaps, the ideal material for its enhanced brightness, exceptionally high quantum yield, narrow emission profile, large Stokes shift, and, most importantly, excellent photostability. Another interesting property associated with qdots is the fluorescent intermittency, or blinking, phenomenon [15], whereby a time trace of fluorescent intensity from a single qdot can switch between two distinct on/off states. This blinking phenomenon is sometimes considered a minor drawback since it might cause temporary trajectory data loss in single-molecule tracking [16]. On the other hand, blinking is often used to identify single molecules, providing a practical benefit [17, 18].

Using qdot-based single-molecule tracking, Dahan and coworkers first reported the diffusion dynamics of individual glycine receptors in living neurons [19]. By conjugating Fab fragments of antibody to single qdots, they were able to track the movement of individual glycine receptors for more than 20 min. Similar approaches were then employed for various protein targets including nerve growth factor (NGF) [20], gamma-aminobutyric acid A receptor ($GABA_A$ R) [21], glycosyl-phosphatidylinositol (GPI)-anchored protein [22], and serotonin transporter (SERT) [23]. In 2008, Murcia et al. pushed the temporal resolution of single-qdot tracking up to an amazing 1,000 frames/s by using an extremely sensitive double micro-channel plate (MCP) image intensifiers, cooled intensified CCD camera (dual MCP ICCD) [24]. Through demonstration of single-qdot tracking in living mice, Tyda et al. further suggested to use single qdots as tumor-targeting nanocarriers for in vivo drug delivery study [25]. Recently, Zhang et al. utilized single qdots as artificial cargo of synaptic vesicles to validate the "kiss-and-run" model of neurotransmitter secretion [17]. Overall, qdot-based single-molecule imaging holds the promise to reveal molecular biological mechanisms well beyond ensemble biochemical techniques.

The process of qdot-based single-molecule microscopy is typically divided into three steps (Fig. 2). The first step is to acquire time-lapse imaging after single target-specific labeling has been made. Single fluorescent molecules should be able to produce diffraction-limited blurred spots in each flame. The second step involves imaging data processing and single-molecule localization

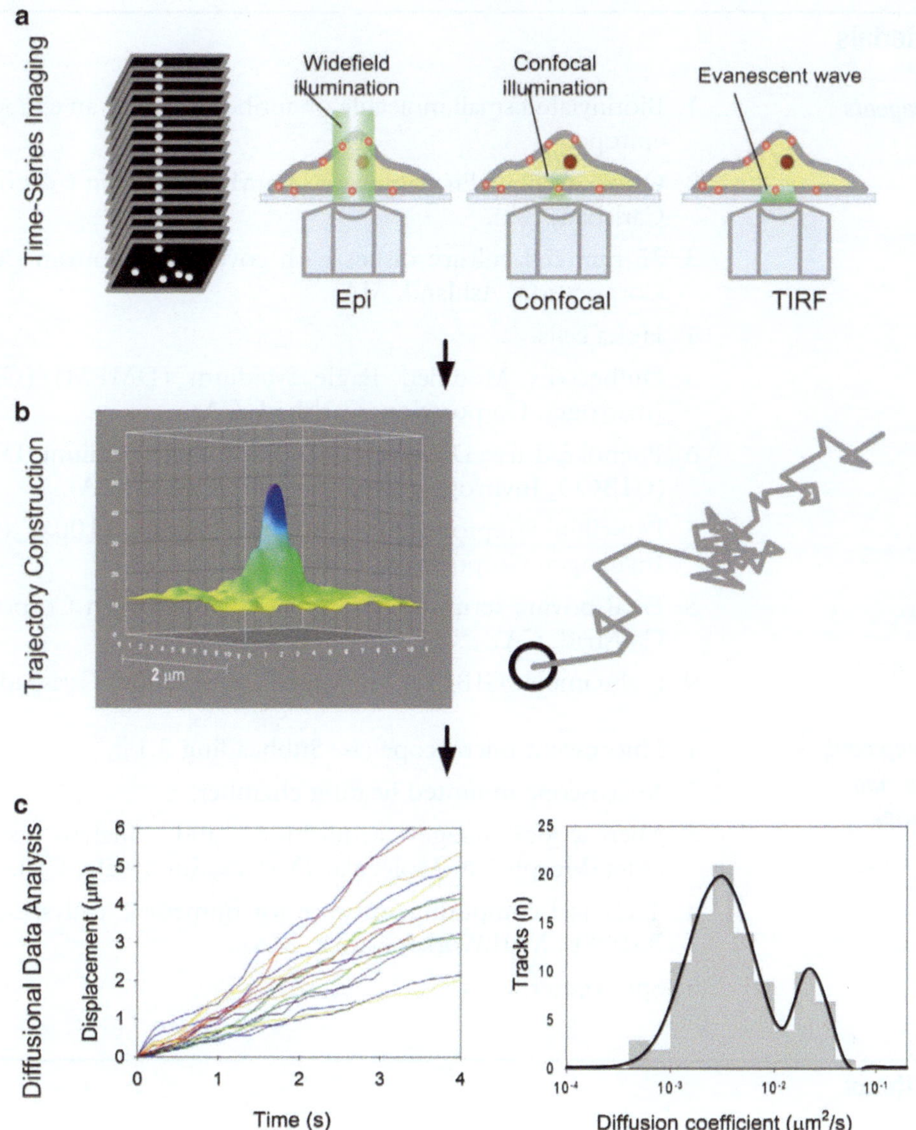

Fig. 2 Approach to single-molecule microscopy. (**a**) Time-lapse images of single fluorophore-tagged biomolecules in living cells acquired from an optical fluorescent microscope system (epifluorescence, confocal, or total internal reflection fluorescence (TIRF) microscope). (**b**) Estimation of the positions of single quantum dots with sub-pixel accuracy is accomplished by fitting the individual spot intensity values with a two-dimensional Gaussian distribution. After the positions of single quantum dots are identified, a trajectory of the target protein (*gray line*) can be subsequently derived from the time-series imaging data. (**c**) The final step is to analyze the diffusional properties (displacement, velocity, and diffusion coefficient) from single-molecule trajectories

from time-lapse images and is therefore normally anticipated as a computationally demanding step. In the third step, the diffusion dynamics can be analyzed from the trajectories of individual molecules, e.g., Brownian motion.

2 Materials

2.1 Reagents

1. Biotinylated small molecule or antibody against an extracellular epitope.
2. Qdot Streptavidin conjugate (1 μM, Invitrogen Corporation, Carlsbad, CA).
3. 35-mm cell culture dishes with cover glass bottom (MatTek Corporation, Ashland, MA).
4. HeLa cells.
5. Dulbecco's Modified Eagle Medium (DMEM) (GIBCO, Invitrogen Corporation, Carlsbad, CA).
6. Phenol red-free Dulbecco's Modified Eagle Medium (DMEM) (GIBCO, Invitrogen Corporation, Carlsbad, CA).
7. Penicillin–Streptomycin Antibiotic Mixture (100×) (Gibco, Invitrogen Corporation, Carlsbad, CA).
8. Fetal bovine serum (FBS) (GIBCO, Invitrogen Corporation, Carlsbad, CA).
9. L-glutamine (GIBCO, Invitrogen Corporation, Carlsbad, CA).

2.2 Equipment, Software, and Accessories

1. Fluorescent microscope (*see* Subheading 3.1).
2. Microscope mounted heating chamber.
3. Microscope image acquisition and analysis software (MetaMorph 7.6, Molecular Devices, Sunnyvale, CA).
4. Technical computing software for numerical analysis (Matlab R2008b, MathWorks, Natick, MA).
5. Spin coater.

3 Methods

3.1 Imaging System Calibration Using Spin-Coated Single Quantum Dots

A general strategy to identify whether the optical system is able to detect individual qdots is to carry out time-lapse imaging of a very dilute qdot solution to avoid multiple qdots overlapping within diffraction-limited distance. Individual qdots are characterized by their blinking properties [8, 18]. The emission intensity of a single qdot is shown in Fig. 3, where a single qdot blinks completely on and off during a time-lapse sequence of 80 s at a 10 Hz frame rate. When the fluorescence is produced by an aggregate structure consisting of several qdots, such blinking effects are completely cancelled out. The protocol described below is based on a custom-built Zeiss Axiovert 200 M inverted fluorescence microscope coupled with a charge-coupled device (CCD) camera (Cool-Snap$_{HQ2}$, Roper Scientific, Trenton, NJ). To track single qdots in real time, the acquisition rate should be set at 10 Hz or higher. However, the

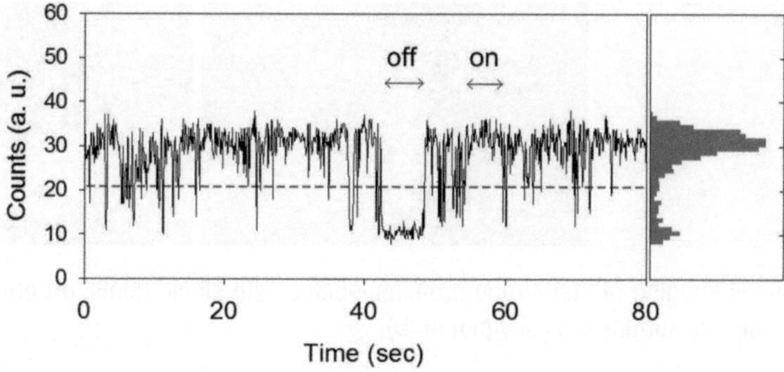

Fig. 3 A typical intensity over time plot from a single blinking qdot. As displayed in the *left panel*, the intensity trajectory of a single qdot displays two dominant states: an "on" state and an "off" state, termed blinking. A predefined intensity threshold is shown by the *dashed line*. *Right panel* displays the probability density distribution

imaging rate is usually limited by the frame readout time of the camera. This particular CCD is chosen due to its adequate 60 % quantum efficiency (QE) throughout the entire visible spectrum range (450–650 mm) with a frame rate >20 at 512 × 512 pixels (*see* **Note 1**). And again, as we introduced in the background section, a more advanced back-illuminated Electron Multiplying CCD (EMCCD) with sub-millisecond temporal resolution will be a much better choice. Imaging should be performed with a high-resolution (63× or 100×) oil-immersion objective lens with numerical aperture of 1.30 or greater.

The sample preparation steps are:

1. Prepare a clean microscope glass slide coverslip (or 35 mm culture dish with a coverslip at the bottom).
2. Add one drop (20 μL) of 100 pM Qdot® ITK™ carboxyl quantum dots solution onto the coverslip.
3. Spin casting the qdot solution on the coverslip for 30 s at 500 rpm (~30 g for common compact spin coater) is sufficient to disperse qdot particles uniformly across the coverslip (*see* **Note 2**).
4. Mount the coverslip on the microscope stage.
5. Qdots are excited using a xenon arc lamp (excitation filter 480/40 BP) and detected with CCD camera thorough appropriate emission filter (600/40 BP for Qdot 605). Acquire time-lapse images (100 ms per frame, 60 s).

3.2 Single-Quantum-Dot Labeling in Living Cells

Single-quantum-dot labeling can be prepared through either a direct labeling (one-step) procedure or an indirect (two-step) protocol. In the direct labeling procedure, the target-specific probe (small-molecule organic ligand, peptide, or antibody) is directly conjugated to the qdot's surface to make ligand–qdot nanoconjugates. Therefore, the cellular labeling strategy could be performed in

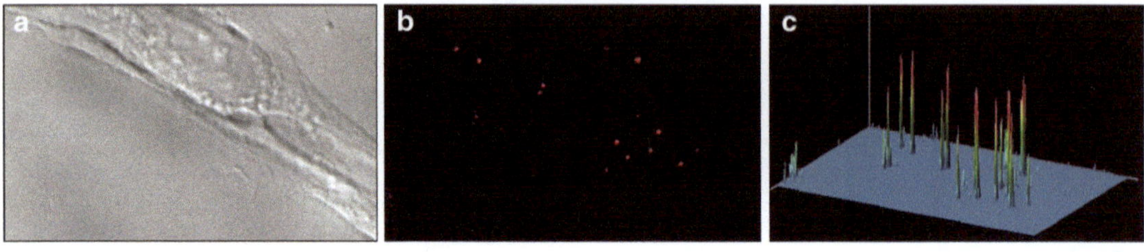

Fig. 4 Example live-cell imaging of membrane proteins labeled with single qdots: (**a**) bright field image, (**b**) fluorescence image, and (**c**) surface intensity plot of (**b**)

one step in which the live cell sample is incubated with a target-specific nanoconjugate prior to fluorescent imaging. In the two-step procedure, the cell sample is first incubated with biotinylated ligand to yield the desired specific ligand–protein binding. After an appropriate washing step, strep-qdots are added as the fluorescent tag of the biotinylated ligand–protein complex for the single-molecule imaging.

We provide a general protocol for single-qdot labeling of adherent cells which is applicable to most mammalian cell lines. The standard protocol given below should be followed:

1. Prepare a 35 mm coverslip-buttoned culture dish with cells that have reached about 50 % confluence.
2. Wash the cells gently three times with phenol red-free culture medium by repeatedly pipetting out.
3. Incubate cells with a biotinylated small-molecule probe (0.5 nM–0.5 μM dependent upon the biological affinity) or antibody (1–10 μg/mL) in red-free DMEM for 20 min at 37 °C. (For one-step labeling protocol, incubate cells with 10–50 pM ligand–qdot nanoconjugates and skip **steps 4** and **5**) (*see* **Note 3**).
4. Wash cells gently three times with phenol red-free culture medium.
5. Incubate the cells with qdot streptavidin conjugate (0.1–0.5 nM) in phenol red-free culture medium for 5 min at 37 °C.
6. Wash the cells at least three times with phenol red-free culture medium.
7. Place the culture dish on the microscope stage with mounted heating chamber and heat to 37 °C.
8. The labeling quality can be observed under fluorescent microscope. Punctate qdot staining should be visible through the eye piece or CCD detector (Fig. 4). Single qdots can be identified by their blinking property.
9. Acquire time-lapse images at 37 °C. In our experiments, acquisition procedure typically lasts for 60 s at 10 Hz rate.

3.3 Tracking Programs for Single-Molecule Analysis

Tracking and trajectory construction is a computationally demanding step of following single-qdot-labeled biological targets through successive images. One of the most important determinants of modern single-molecule tracking techniques is the nanometer accuracy, which is heavily weighted by the PSF fitting to localize the centroid position for sub-pixel resolution, normally demonstrated as a fitting of a 2D Gaussian function to a PSF. For practical application, estimated background-corrected intensities of an image are normally filtered out, a necessary step for the calculation of centroid position (x_0, y_0). After locations of single molecules are identified in each frame, the next step is to link the detected single-molecule positions. However, single-qdot blinking brings additional difficulties for the trajectory generation, as the spots can temporarily disappear. A practical and most frequently used approach is to define a tolerance limit of blinking frames (usually ≤10 frames) and process an additional association step in the trajectory generation algorithm to merge multiple trajectories into one [25]. This procedure allows tracking to continue and thus compensates for the transient data loss caused by qdot blinking. In addition, qdot blinking frequency is dependent on the excitation power; hence, it is generally recommended to perform single-qdot tracking experiments with low-power excitation if signal intensity is sufficiently high.

An ImageJ plug-in for single-molecule/particle tracking offers several user-friendly features including an easily understandable interface, free online tutorial, and computationally efficient process. The program is free to download at ImageJ website: http://rsbweb.nih.gov/ij/plugins/index.html (*see* **Note 4**). In addition, particle tracking using IDL, developed by Crocker and Grier, provides a total solution including 2D Gaussian fit for spot localization, trajectory generation, as well as MSD calculation. The algorithms with detailed tutorial are freely available at http://www.physics.emory.edu/~weeks/idl/index.html. Matlab version of these routines can be found at http://physics.georgetown.edu/matlab/ (Fig. 5). With recent advances in computing power and numerical software, the development of tracking algorithms has evolved rapidly during the past few years in supporting better correspondence for motion detection [27], high computational efficiency [28], or 3-D motion segmentation and localization [29]. All the tracking algorithms mentioned above may, however, require technical training to operate since they all established under technical computing environments such as Matlab, IDL, or C++.

Below is a general tracking procedure using ParticleTracker:

1. Use the *File/Open* command in the ImageJ to import the prerecorded TIF stack or uncompressed avi file (Fig. 6a). If the avi file contains multichannel imaging data, use the *Image/Color/RGB Split* to qdot data channel extraction.

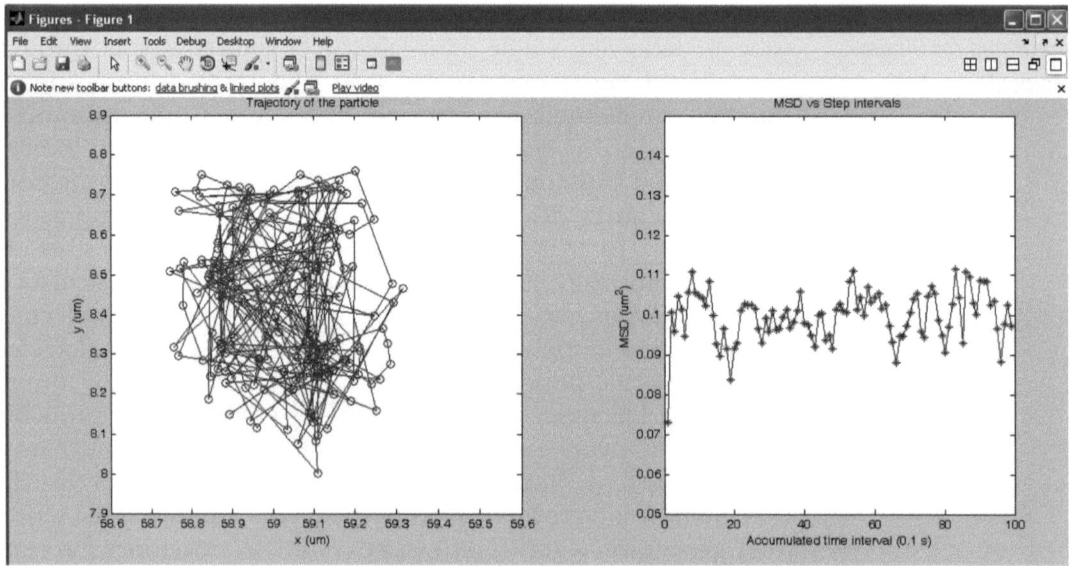

Fig. 5 Snapshot of the interface of Matlab-based particle tracking program originally developed from *particle tracking using IDL* algorithm. Data obtained from 600 frames of single-qdot imaging. *Left panel* indicates the 2D trajectory and *right panel* shows the MSD over time

2. Next, click the *Plug-ins/Particle Detector* and *Tracker* command. If RGB images or images with greater than 8 bits are loaded for tracking, a checkable menu item will show up to ask whether the images are converted to 8 bits. If running on a computer with fewer than 2 GB of memory installed, it is strongly recommended to convert to 8 bits to reduce the memory consumption.
3. As indicated in Fig. 6b, three basic parameters for particle detection are given. *Radius*: Approximate radius of the particles in the images in units of pixels. *Cutoff*: The score cutoff for the non-particle discrimination. *Percentile*: The percentile (r) that determines which bright pixels are accepted as particles. Click on preview detected and then the successfully detected spots will be circulated. Here we recommend to use *radius* = 3, *cutoff* = 0, and *percentile* = 0.1 as initial guess, but these values might vary based on the images. Start with our recommended parameter and change these values until most of the visible particles are detected after clicking the preview button.
4. After setting the parameters for the detection, set up the particle linking parameters (*Displacement* and *Link Range*) in the bottom of the dialog window (Fig. 6b). Here the *Displacement* parameter means the maximum number of pixels a particle is allowed to move between two succeeding frames. The *Link Range* parameter is used to specify the number of subsequent frames that is taken into account to determine the optimal correspondence matching. We recommend to use *Displacement* = 2 and *Link Range* = 10 as initial guess, and again, these parameters can also be modified after viewing the initial results.

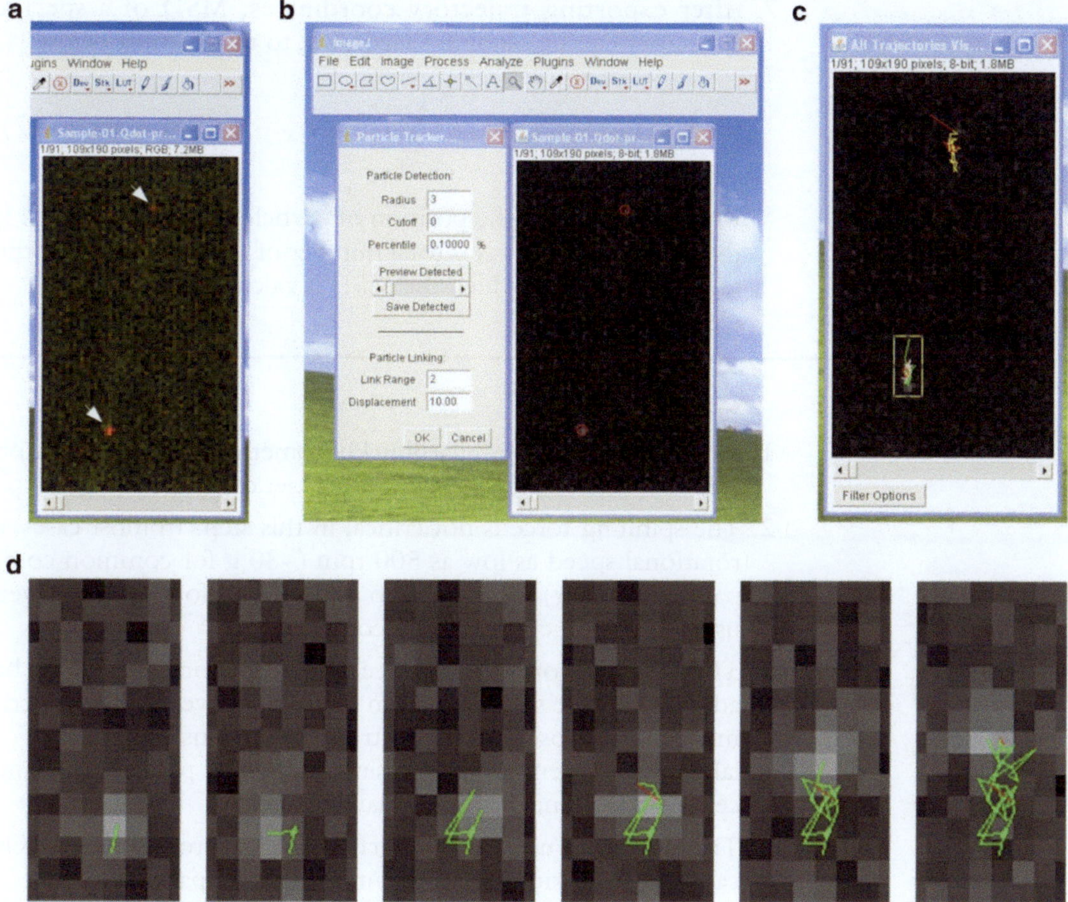

Fig. 6 Steps of single-qdot tracking using ParticleTracker—an ImageJ plug-in. (**a**) A typical raw frame from a time-lapse single-qdot-labeling movie. *White arrows* indicate the qdot-labeled target proteins. (**b**) Image conversion and preview detection from the raw frame. Image conversion to Gray 8 is preferred to increase computational efficiency. *Red circle* masks the successfully targeted spots for tracking after executing the Preview Detected function. (**c**) Visualization of all trajectories after executing the Show Detected function. Particular area of interest can be selected and zoomed in as indicated in *yellow box*. (**d**) Time-lapse trajectory from the selected area of interest. *Red line* drawn indicates the "gaps" in the trajectory

5. Next push the OK button and the result window will then be displayed (Fig. 6c) after seconds to minutes computational calculation (for 600 frames of 128 × 128 pixel images takes less than 1 min on a regular dual core PC). With the *Filter Options* button given on the dialog window, you can filter out trajectories under a given length. Particular trajectory of interest can be selected by clicking it once with the mouse left button (*see* yellow box of Fig. 6c).

6. The selected trajectory can be displayed in a separate window by clicking on the *Focus on Selected Trajectory* button. The visualization of the selected trajectory can then be saved individually in .gif format (Fig. 6d). Detected time-series trajectory coordinates can also be exported in a single .txt file for further analyses.

7. After exporting trajectory coordinates, MSD of a specific trajectory can be obtained according to the formula below:

$$\text{MSD}(n\Delta t) = (N-n)^{-1}\sum_{i=1}^{N-n}[(x_{i+n}-x_i)^2+(y_{i+n}-y_i)^2], \tag{7}$$

where x_i and y_i are the position of particle on the frame i, Δt is the time resolution, N is total number of frames, and $n\Delta t$ is the time interval over which the MSD is calculated.

4 Notes

1. Detailed information regarding Photometrics CCD specification can be found at http://www.photomet.com.
2. The spinning force is not critical in this step. In most cases, a rotational speed as low as 500 rpm (~30 g for common compact spin coater) is sufficient to achieve a uniform spread when using common compact spin coater.
3. The labeling concentration/cell type relationship should be adjusted for the surface protein expression level. In our experiments, we choose low concentrations for transfected cells. For labeling endogenously expressing membrane proteins in living cells, higher concentrations may be needed.
4. The algorithm used in the ParticleTracker program can easily cause false linking of different molecules/particles between frames. This could lead to incorrect trajectory construction. It may be improved by manual relinking with visual inspection. However, potential problems and limitations can still be associated with such manual relinking. Due to its respectable efficiency, ParticleTracker program is suitable for preliminary screening tests. For serious and in-depth analysis, we recommend to perform particle tracking using IDL or the Matlab-based tracking routines.

Acknowledgments

The authors thank Drs. David Piston and Sam Wells for helpful advice with single-quantum-dot tracking experimental setup. We thank colleagues in the group, especially to Dr. James McBride and Oleg Kovtun, for helpful discussions and suggestions. This work was supported by grants from National Institutes of Health (R01EB003778).

References

1. Dunn RC (1999) Near-field scanning optical microscopy. Chem Rev 99:2891–2928
2. Gordon MP, Ha T, Selvin PR (2004) Single-molecule high-resolution imaging with photobleaching. Proc Natl Acad Sci USA 101: 6462–6465
3. Willig KI, Rizzoli SO, Westphal V, Jahn R, Hell SW (2006) STED microscopy reveals that synaptotagmin remains clustered after synaptic vesicle exocytosis. Nature 440:935–939
4. Huang B, Wang WQ, Bates M, Zhuang XW (2008) Three-dimensional super-resolution imaging by stochastic optical reconstruction microscopy. Science 319:810–813
5. Zhang B, Zerubia J, Olivo-Marin JC (2007) Gaussian approximations of fluorescence microscope point-spread function models. Appl Opt 46:1819–1829
6. Thomann D, Rines DR, Sorger PK, Danuser G (2002) Automatic fluorescent tag detection in 3D with super-resolution: application to the analysis of chromosome movement. J Microsc 208:49–64
7. Cheezum MK, Walker WF, Guilford WH (2001) Quantitative comparison of algorithms for tracking single fluorescent particles. Biophys J 81:2378–2388
8. Kubitscheck U, Kückmann O, Kues T, Peters R (2000) Imaging and tracking of single GFP molecules in solution. Biophys J 78: 2170–2179
9. Alivisatos P (2004) The use of nanocrystals in biological detection. Nat Biotechnol 22:47–52
10. Chan WC, Nie S (1998) Quantum dot bioconjugates for ultrasensitive nonisotopic detection. Science 281:2016–2018
11. Bruchez M, Moronne M, Gin P, Weiss S, Alivisatos AP (1998) Semiconductor nanocrystals as fluorescent biological labels. Science 281:2013–2016
12. Rosenthal SJ, Tomlinson I, Adkins EM, Schroeter S, Adams S, Swafford L, McBride J, Wang Y, DeFelice LJ, Blakely RD (2002) Targeting cell surface receptors with ligand-conjugated nanocrystals. J Am Chem Soc 124:4586–4594
13. Kim S, Lim YT, Soltesz EG, De Grand AM, Lee J, Nakayama A, Parker JA, Mihaljevic T, Laurence RG, Dor DM, Cohn LH, Bawendi MG, Frangioni JV (2004) Near-infrared fluorescent type II quantum dots for sentinel lymph node mapping. Nat Biotechnol 22:93–97
14. Chang JC, Tomlinson ID, Warnement MR, Iwamoto H, DeFelice LJ, Blakely RD, Rosenthal SJ (2011) A fluorescence displacement assay for antidepressant drug discovery based on ligand-conjugated quantum dots. J Am Chem Soc 133:17528–17531
15. Nirmal M, Dabbousi BO, Bawendi MG, Macklin JJ, Trautman JK, Harris TD, Brus LE (1996) Fluorescence intermittency in single cadmium selenide nanocrystals. Nature 383:802–804
16. Wang X, Ren X, Kahen K, Hahn MA, Rajeswaran M, Maccagnano-Zacher S, Silcox J, Cragg GE, Efros AL, Krauss TD (2009) Non-blinking semiconductor nanocrystals. Nature 459:686–689
17. Zhang Q, Li Y, Tsien RW (2009) The dynamic control of kiss-and-run and vesicular reuse probed with single nanoparticles. Science 323: 1448–1453
18. Thompson MA, Lew MD, Badieirostami M, Moerner WE (2009) Localizing and tracking single nanoscale emitters in three dimensions with high spatiotemporal resolution using a double-helix point spread function. Nano Lett 10:211–218
19. Dahan M, Levi S, Luccardini C, Rostaing P, Riveau B, Triller A (2003) Diffusion dynamics of glycine receptors revealed by single-quantum dot tracking. Science 302:442–445
20. Cui B, Wu C, Chen L, Ramirez A, Bearer EL, Li W-P, Mobley WC, Chu S (2007) One at a time, live tracking of NGF axonal transport using quantum dots. Proc Natl Acad Sci USA 104:13666–13671
21. Bouzigues C, Morel M, Triller A, Dahan M (2007) Asymmetric redistribution of GABA receptors during GABA gradient sensing by nerve growth cones analyzed by single quantum dot imaging. Proc Natl Acad Sci USA 104:11251–11256
22. Fabien P, Xavier M, Gopal I, Emmanuel M, Hsiao-Ping M, Shimon W (2009) Dynamic partitioning of a glycosyl-phosphatidylinositol-anchored protein in glycosphingolipid-rich microdomains imaged by single-quantum dot tracking. Traffic 10:691–712
23. Chang JC, Tomlinson ID, Warnement MR, Ustione A, Carneiro AM, Piston DW, Blakely RD, Rosenthal SJ (2012) Single molecule analysis of serotonin transporter regulation using antagonist-conjugated quantum dots reveals restricted, p38 MAPK-dependent mobilization underlying uptake activation. J Neurosci 32:8919–8929
24. Murcia MJ, Minner DE, Mustata G-M, Ritchie K, Naumann CA (2008) Design of quantum dot-conjugated lipids for long-term, high-speed tracking experiments on cell surfaces. J Am Chem Soc 130:15054–15062

25. Tada H, Higuchi H, Wanatabe TM, Ohuchi N (2007) In vivo real-time tracking of single quantum dots conjugated with monoclonal anti-HER2 antibody in tumors of mice. Cancer Res 67:1138–1144
26. Ehrensperger M-V, Hanus C, Vannier C, Triller A, Dahan M (2007) Multiple association states between glycine receptors and gephyrin identified by SPT analysis. Biophys J 92:3706–3718
27. Jaqaman K, Loerke D, Mettlen M, Kuwata H, Grinstein S, Schmid SL, Danuser G (2008) Robust single-particle tracking in live-cell time-lapse sequences. Nat Methods 5:695–702
28. Smith CS, Joseph N, Rieger B, Lidke KA (2010) Fast, single-molecule localization that achieves theoretically minimum uncertainty. Nat Methods 7:373–375
29. Ram S, Prabhat P, Chao J, Sally Ward E, Ober RJ (2008) High accuracy 3D quantum dot tracking with multifocal plane microscopy for the study of fast intracellular dynamics in live cells. Biophys J 95:6025–6043

Chapter 7

Three-Dimensional Molecular Imaging with Photothermal Optical Coherence Tomography

Melissa C. Skala, Matthew J. Crow, Adam Wax, and Joseph A. Izatt

Abstract

Optical coherence tomography (OCT) is a three-dimensional optical imaging technique that has been successfully implemented in ophthalmology for imaging the human retina, and in studying animal models of disease. OCT can nondestructively visualize structural features in tissue at cellular-level resolution, and can exploit contrast agents to achieve molecular contrast. Photothermal OCT relies on the heat-producing capabilities of antibody-conjugated gold nanoparticles to achieve molecular contrast. A pump laser at the nanoparticle resonance wavelength is used to heat the nanoparticles in the sample, and the resulting changes in the index of refraction around the nanoparticles are detected by phase-sensitive OCT. Lock-in detection of the pump beam amplitude-modulated frequency and the detector frequency allow for high-sensitivity images of molecular targets. This approach is attractive for nondestructive three-dimensional molecular imaging deep (approximately 2 mm) within biological samples. The protocols described here achieve a sensitivity of 14 parts per million (weight/weight) nanoparticles in the sample, which is sufficient to differentiate EGFR (epidermal growth factor receptor)-overexpressing cells from minimally expressing cells in three-dimensional cell constructs.

Key words Nanospheres, Microscopy, Optical coherence tomography, Gold, Contrast media, Epidermal growth factor receptor, Fourier analysis, Equipment design, MDA-MB-435, MDA-MB-468

1 Introduction

Molecular imaging is a powerful tool for investigating biological signaling, disease processes, and potential therapies in both in vivo and in vitro systems. Microscopy, including confocal and multiphoton microscopy, has been the standard for high-resolution molecular imaging in live cells and tissues. However, these microscopy techniques suffer from relatively shallow imaging depths. MRI and PET have been the standard for functional imaging deep within the body, with the caveat of relatively poor resolution. Optical coherence tomography (OCT) fills a niche between high-resolution

Sandra J. Rosenthal and David W. Wright (eds.), *NanoBiotechnology Protocols*, Methods in Molecular Biology, vol. 1026,
DOI 10.1007/978-1-62703-468-5_7, © Springer Science+Business Media New York 2013

microscopy techniques and whole-body imaging techniques with relatively good resolution (~1–10 μm) and penetration depths (~1–2 mm) in tissue. Molecular imaging in this regime would be a powerful tool for scientists and clinicians. Two examples with potentially high impact are imaging the effects of antiangiogenic treatment in age-related macular degeneration, and whole-tumor imaging of molecular microenvironments.

OCT is intrinsically insensitive to incoherent scattering processes such as fluorescence and spontaneous Raman scattering, which are central to optical molecular imaging, because OCT depends on coherent detection of scattered light. Gold nanoparticles are attractive contrast agents for OCT because they are biocompatible, do not exhibit photobleaching or cytotoxicity, and are tunable through a broad range of wavelengths including the visible and near-infrared regions. Gold nanoparticles currently under development for molecular contrast OCT include highly scattering gold nanoshells [1], gold nanocages [2], and gold nanorods [3]. Gold nanoshells are also under development as photothermal contrast agents [4].

Photothermal imaging [5–7] provides one potential method for increasing the molecular contrast of OCT over a highly scattering background. In photothermal imaging, strong optical absorption of a small metal particle at its plasmon resonance results in a change in temperature around the particle (photothermal effect). This temperature change leads to a variation in the local index of refraction that can be optically detected with an amplitude-modulated heating beam that spatially overlaps with the focus position of the sample arm of an interferometer. Previous work has shown that photothermal interference contrast images of gold nanoparticles from a modified DIC microscope are insensitive to a highly scattering background [5]. We have applied this concept to OCT with the added benefits of depth resolution and increased imaging depth. Our photothermal OCT system has a measured sensitivity of 14 parts per million (ppm, weight/weight), and we have used this system to measure epidermal growth factor receptor (EGFR) expression from live monolayers of cells and in three-dimensional tissue constructs [8]. Protocols for nanoparticle conjugation, labeling of cell monolayers and three-dimensional cell constructs, dark-field microspectral imaging, and photothermal OCT imaging follow.

2 Materials

2.1 Cell Culture

1. Cells that overexpress EGFR (MDA-MB-468) [9] and cells that express low levels of EGFR (MDA-MB-435) [10] were obtained from the American Type Culture Collection (ATCC).
2. Dulbecco's Modified Eagle's Medium (DMEM) (Gibco/BRL, Bethesda, MD) supplemented with 10 % fetal bovine serum (FBS, HyClone, Ogden, UT).

3. Chambered coverglasses (Laboratory-Tek).
4. Low-gelling point agarose (Sigma-Aldrich Co.).
5. 24-well inserts (6.5 mm diameter, Transwell, Fisher Scientific).
6. 10 % dimethyl sulfoxide (DMSO).

2.2 Gold Nanosphere Antibody Conjugation

1. Anti-epidermal growth factor receptor (anti-EGFR) mAb (E2156, Sigma, ~1.5 mg/mL).
2. HEPES buffer.
3. Gold colloids of 60 nm diameter (Ted Pella, Inc.).
4. Polyethylene glycol (PEG) compound (Sigma P2263).

2.3 Imaging Instrumentation

1. Dark-field microspectroscopy system [11, 12].
   (a) Inverted microscope (Axiovert 200, Carl Zeiss, Inc.).
   (b) Color camera (CoolSnap cf, Photometrics).
   (c) Line-imaging spectrometer (SpectraPro 2150i, Action Research).
2. Photothermal optical coherence tomography (OCT) system (*see* **Note 1**) [8].
   (a) Diode pumped solid-state frequency doubled Nd:YAG laser as the heating laser (Coherent, Verdi; *see* **Note 2**).
   (b) Super luminescent diode light source (Superlum) centered at 840 nm with a full width at half-maximum bandwidth of 52 nm (resulting in 6 μm axial resolution).
   (c) 30 mm focal length focusing lens (20 μm imaging spot size on the sample; *see* **Note 3**).
   (d) Two-dimensional scanning mirrors (galvos, Cambridge Technology).
   (e) Custom spectrometer with 1,024 pixel line-scan CCD camera (ATMEL, Aviiva).
   (f) Software to control lateral scanning, perform data acquisition, rescaling from wavelength to wavenumber, Fourier Transform, two-dimensional B-scan display, and data archiving in real time (Bioptigen Inc.).
   (g) Optical chopper to amplitude-modulate the heating laser (Thorlabs).

3 Methods

3.1 Gold Nanosphere Antibody Conjugation [11–14]

1. Dilute 1 mL of the 60 nm diameter gold colloid with 125 mL 20 μM HEPES buffer.
2. Dilute 30 μL anti-EGFR mAb in 20 mM HEPES buffer to prepare a 3 % (volume/volume) anti-EGFR solution.

3. Adjust the pH of the colloid and antibody preparations to 7.0 ± 0.2 by the addition of 100 nM K_2CO_3.
4. Mix the pH-adjusted colloid and antibody preparations and allow to conjugate at room temperature for 20 min on an oscillator operating at 190 cycles/min.
5. After verifying antibody attachment (*see* **Note 4**), add 200 μL of 1 % PEG compound to the remainder of the conjugated nanoparticle suspension and allow the solution to interact at room temperature for 10 min.
6. At the end of the interaction period, centrifuge the solution at $2245 \times g$ until a pellet is formed (~10 min) (*see* **Note 5**).
7. Withdraw the supernatant and resuspend the nanoparticle pellet in 1 mL of phosphate buffered saline (*see* **Note 6**).

3.2 Cell Monolayer Experiments

1. Suspend 80,000 cells in 1 mL of media, plate onto 2.0 mL chambered cover glasses, and incubate for 12–16 h to allow for cell adhesion.
2. Exchange media with a mixture of 0.5 mL antibody-conjugated nanosphere suspension (2.35×10^{10} nanospheres/mL; *see* **Note 7**) and 0.5 mL fresh media.
3. Incubate at 37°°C for 20 min, remove the media, rinse cells twice with fresh media, and add fresh media once more for imaging immediately afterward (*see* **Note 8**).
4. Image labeled monolayers using dark-field microspectroscopy and photothermal OCT (*see* **Note 9** and Fig. 1).

3.3 Dark-Field Microspectroscopy

1. Acquire color images of cells to confirm spatial distribution nanoparticle labeling.
2. Acquire dark-field scattering spectra from the labeled cells to confirm plasmon resonance peak of the functionalized nanoparticles.

3.4 Photothermal OCT Imaging

1. Acquire multiple OCT depth scans (A-scans) at each position in the sample.
2. Calculate the photothermal signal at each point in the cross-sectional image.
   (a) Take the Fourier transform of the phase vs. time at each point in the image.
   (b) The photothermal signal is the peak at the pump laser frequency, minus the background (Fig. 2).

3.5 Three-Dimensional Tissue Culture

1. Mix 3 % low-gelling point agarose with a cell suspension in media (100×10^6 cells/mL) to obtain a homogeneous cell distribution.
2. Pour agarose/cell/media mixture into 24-well inserts at 2–3 mm thickness for gelation at room temperature (*see* **Note 10**).

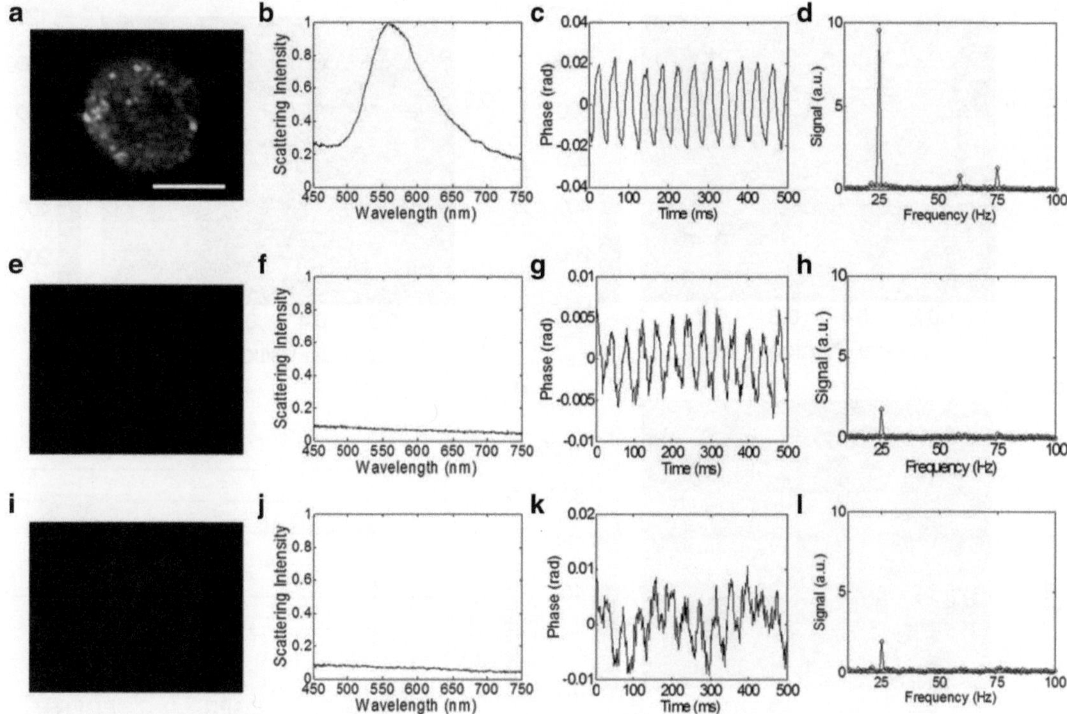

Fig. 1 EGFR expression and nanoparticle labeling was confirmed in EGFR+/nanosphere+ cells (**a**–**d**), with two control groups that include EGFR–/nanosphere+ (**e**–**h**) and EGFR+/nanosphere– (**i**–**l**) using dark-field microscopy (*panels* **a**, **e**, **i**) and microspectroscopy (*panels* **b**, **f**, **j**). The phase as a function of time in the photothermal system is plotted for the experimental (*panel* **c**) and control groups (*panels* **g**, **k**), and the Fourier transform of the phase confirms oscillations at 25 Hz (the pump laser modulation frequency) (*panels* **d**, **h**, **l**). In three repeated experiments, the photothermal signal from overexpressing cell monolayers was at least 9 dB higher than the highest signal from the cells that express low levels of EGFR. The repeated experiments were performed at pump powers of 7.5–8.5 kW/cm^2. Scale bar is 20 μm. Reproduced with permission from ref. [8]

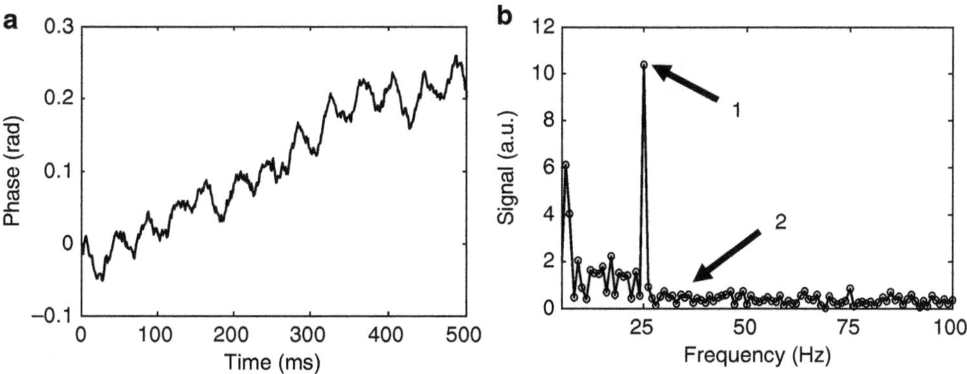

Fig. 2 Phase of the tissue-like phantom (polystyrene spheres with $\mu_s = 100$ cm^{-1}) with 84 ppm nanospheres and 25 Hz pump frequency (**a**). The definition of the photothermal signal (**b**) in the Fourier-transformed phase is the peak at 25 Hz (*arrow 1*, the pump beam modulation frequency in these experiments) minus the background (*arrow 2*, 27–50 Hz). Reproduced with permission from ref. [8]

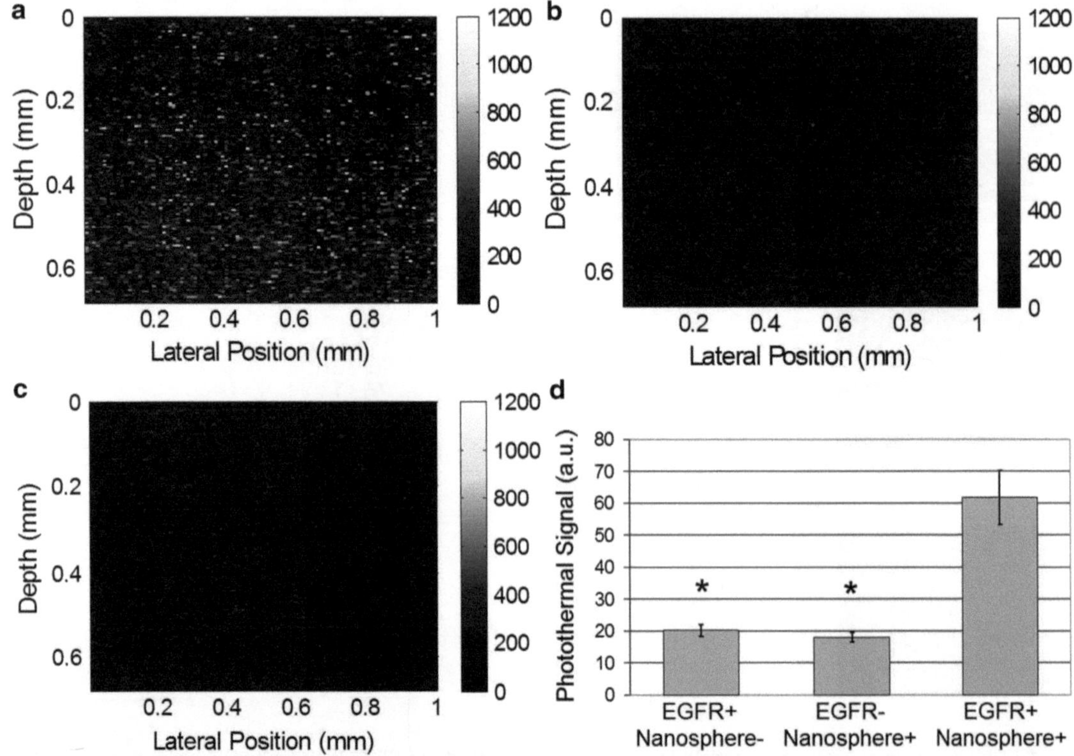

Fig. 3 Images of EGFR expression in three-dimensional cell constructs containing EGFR+ cells (MDA-MB-468) with and without antibody-conjugated nanospheres (**a** and **c**, respectively) and EGFR− cells (MDA-MB-435) with antibody-conjugated nanospheres (**b**). There was a significant increase in the photothermal signal from EGFR-overexpressing cell constructs labeled with antibody-conjugated nanospheres (**d**) compared to the two controls (EGFR+/Nanosphere− and EGFR−/Nanosphere+). $N = 17$ images for each group, (*, $p < 0.0001$). Pump power 8.5 kW/cm^2. Reproduced with permission from ref. [8]

3. After gelation, topically label cell constructs with antibody-conjugated nanospheres in the same manner as the monolayer experiment (*see* Subheading 3.1), except add 10 % DMSO to the antibody-conjugated nanosphere suspension to allow the solution to penetrate the three-dimensional construct, and incubate for 30 min [10].
4. Wash construct two times with saline.
5. Place a coverslip on the construct for photothermal OCT imaging.
6. Photothermal OCT imaging parameters included 1,000 sequential A-scans (1 ms integration time for each A-scan) at each of 110 lateral positions across the top of the construct (1 mm scan length in the lateral dimension). *See* Fig. 3.

4 Notes

1. Commercial spectral domain OCT systems can be purchased from a variety of sources (Thorlabs, Bioptigen, Inc., etc.), and may be used with the heating laser and chopper to perform photothermal OCT measurements.
2. The pump laser wavelength must overlap with the resonance peak of the nanoparticle of interest, and must provide sufficient power to produce a measurable photothermal signal. In these experiments, 20 mW of 532 nm light was incident on the sample.
3. The focused spot of the pump beam overlaps with the focused spot of the imaging beam (20 μm pump beam spot size on the sample).
4. Verify antibody attachment by removing 200 μL of the resulting conjugated colloid and mix with 10 μL of 10 % NaCl. It is well known that the addition of NaCl will cause nanoparticle aggregation [15], resulting in a color change to the solution, unless the nanoparticle surface has been well coated.
5. Centrifugation step can be repeated if necessary to form a pellet.
6. Resuspend in 0.5 mL of phosphate buffered saline twice to ensure complete removal of the excess PEG.
7. The antibody-conjugated nanosphere suspension of 2.35×10^{10} nanospheres/mL corresponds to 1.3 optical density in a 1 cm path-length cuvette.
8. For photothermal OCT imaging of cell monolayers, replace the media (which contains phenol red that could interfere with the photothermal experiments) with saline.
9. Confirm viability of cells after photothermal imaging using trypan blue exclusion. Our previous experiments show no loss of cell viability due to photothermal imaging.
10. The viability of the cells in the three-dimensional construct can be verified by applying a live-dead fluorescence stain (Invitrogen) without nanosphere labeling. Confocal microscopy can be used to quantify the ratio of live (green fluorescence) to dead (red fluorescence) cells.

References

1. Agrawal A, Huang S, Wei Haw Lin A et al (2006) Quantitative evaluation of optical coherence tomography signal enhancement with gold nanoshells. J Biomed Opt 11: 041121
2. Cang H, Sun T, Li ZY et al (2005) Gold nanocages as contrast agents for spectroscopic optical coherence tomography. Opt Lett 30: 3048–3050
3. Oldenburg AL, Hansen MN, Zweifel DA, Wei A, Boppart SA (2006) Plasmon-resonant gold nanorods as low backscattering albedo contrast agents for optical coherence tomography. Opt Express 14(15):6724–6738

4. Adler DC, Huang S-W, Huber R, Fujimoto JG (2008) Photothermal detection of gold nanoparticles using phase-sensitive optical coherence tomography. Opt Express 16: 4376–4393
5. Boyer D, Tamarat P, Maali A, Lounis B, Orrit M (2002) Photothermal imaging of nanometer-sized metal particles among scatterers. Science 297:1160–1163
6. Telenkov SA, Dave DP, Sethuraman S, Akkin T, Milner TE (2004) Differential phase optical coherence probe for depth-resolved detection of photothermal response in tissue. Phys Med Biol 49:111–119
7. Zharov VP, Galanzha EI, Tuchin VV (2005) Integrated photothermal flow cytometry in vivo. J Biomed Opt 10:051502
8. Skala MC, Crow MJ, Wax A, Izatt JA (2008) Photothermal optical coherence tomography of epidermal growth factor receptor in live cells using immunotargeted gold nanospheres. Nano Lett 8:3461–3467
9. Bailey KE, Costantini DL, Cai Z et al (2007) Epidermal growth factor receptor inhibition modulates the nuclear localization and cytotoxicity of the Auger electron emitting radiopharmaceutical ^{111}In-DTPA human epidermal growth factor. J Nucl Med 48:1562–1570
10. Aaron J, Nitin N, Travis K et al (2007) Plasmon resonance coupling of metal nanoparticles for molecular imaging of carcinogenesis in vivo. J Biomed Opt 12:034007
11. Curry A, Hwang WL, Wax A (2006) Epi-illumination through the microscope objective applied to darkfield imaging and microspectroscopy of nanoparticle interaction with cells in culture. Opt Express 14:6535–6542
12. Curry AC, Crow M, Wax A (2008) Molecular imaging of epidermal growth factor receptor in live cells with refractive index sensitivity using dark-field microspectroscopy and immunotargeted nanoparticles. J Biomed Opt 13:014022
13. El-Sayed IH, Huang X, El-Sayed MA (2005) Surface plasmon resonance scattering and absorption of anti-EGFR antibody conjugated gold nanoparticles in cancer diagnostics: applications in oral cancer. Nano Lett 5:829–834
14. Sokolov K, Follen M, Aaron J et al (2003) Real-time vital optical imaging of precancer using anti-epidermal growth factor receptor antibodies conjugated to gold nanoparticles. Cancer Res 63:1999–2004
15. Sato K, Hosokawa K, Maeda M (2003) Rapid aggregation of gold nanoparticles induced by non-cross-linking DNA hybridization. J Am Chem Soc 125:8102–8103

Chapter 8

Detecting Respiratory Syncytial Virus Using Nanoparticle-Amplified Immuno-PCR

Jonas W. Perez, Nicholas M. Adams, Grant R. Zimmerman, Frederick R. Haselton, and David W. Wright

Abstract

Early-stage detection is essential for effective treatment of pediatric virus infections. In traditional immuno-PCR, a single antibody recognition event is associated with one to three DNA tags, which are subsequently amplified by PCR. In this protocol, we describe a nanoparticle-amplified immuno-PCR assay that combines antibody recognition of traditional ELISA with a 50-fold nanoparticle valence amplification step followed by amplification by traditional PCR. The assay detects a respiratory syncytial virus (RSV) surface fusion protein using a Synagis antibody bound to a 15 nm gold nanoparticle co-functionalized with thiolated DNA complementary to a hybridized 76-base Tag DNA. The Tag DNA to Synagis ratio is 50 to 1. The presence of virus particles triggers the formation of a "sandwich" complex comprised of the gold nanoparticle construct, virus, and a 1 μm antibody-functionalized magnetic particle used for extraction. Virus-containing complexes are isolated using a magnet, DNA tags released by heating to 95 °C, and detected via real-time PCR. The limit of detection of the nanoparticle-amplified immuno-PCR assay was compared to traditional ELISA and traditional RT-PCR using RSV-infected HEp-2 cell extracts. Nanoparticle-amplified immuno-PCR showed a ~4,000-fold improvement in the limit of detection compared to ELISA and a fourfold improvement in the limit of detection compared to traditional RT-PCR. Nanoparticle-amplified immuno-PCR offers a viable platform for the development of an early-stage diagnostics requiring an exceptionally low limit of detection.

Key words Immuno-PCR, PCR, Viral detection, Respiratory syncytial virus, Gold nanoparticle, Magnetic extraction

1 Introduction

The most important cause of severe respiratory illness in young children and infants is respiratory syncytial virus (RSV). RSV is the major cause of infantile bronchiolitis and the most frequent cause of hospitalization of infants and young children in industrialized countries [1]. In the United States, RSV is estimated to account for over 125,000 infant hospitalizations for bronchiolitis or pneumonia per year [2]. Although not fatal to the majority of patients, RSV can be deadly in immunocompromised populations, and in

Sandra J. Rosenthal and David W. Wright (eds.), *NanoBiotechnology Protocols*, Methods in Molecular Biology, vol. 1026, DOI 10.1007/978-1-62703-468-5_8, © Springer Science+Business Media New York 2013

the USA, RSV is responsible for 10,000 deaths annually in patients over the age of 65 [3, 4]. Additionally, naturally acquired immunity to RSV is not complete, and recurrent infection can be frequent [5–7].

Early attempts at developing a formalin-inactivated RSV vaccine were hindered by an increase in disease severity upon subsequent infection of vaccinated patients [8–10]. Because of such problems, there is no approved vaccine for RSV [11]. Without a vaccine, treatment is typically limited to prophylactic passive immunization with neutralizing humanized monoclonal antibodies (palivizumab) or antivirals (ribavirin) [5, 12]. The cost associated with monthly prophylactic passive immunization is expensive, so it typically is only administered to premature or other high-risk infants [12, 13]. Another option, antivirals, is only effective when used early in the course of infection, making early detection crucial [14].

To date, one of the most sensitive diagnostics available for the detection of RSV has been the use of reverse transcription polymerase chain reaction (RT-PCR) to amplify and detect RSV genetic material [15, 16]. However, detection of the negative-sense RNA genome or mRNA of the virus comes with a few drawbacks. Firstly, RNA is very sensitive to degradation by RNases and typically has a shorter half-life than proteins [17, 18]. This, in turn, places increased importance on preparation, pre-assay handling, and storage of samples for RT-PCR diagnostics [19]. Secondly, studies have shown that RSV mRNA expressed during infection is cyclical and can limit the amount of mRNA present to detect at any given time; however, the surface fusion protein (F-protein) increases monotonically with time [20]. Further, the link between RT-PCR viral load and disease severity has yet to be established [16]. In contrast, an immuno-PCR (IPCR)-based diagnostic for the detection of RSV comes with two inherent strengths. Firstly, empty viral nucleocapsids can be detected because only the presence of a specific protein is required for a positive result and even fragments of cell wall that have the protein of interest expressed can be detected [21]. Secondly, each virion may contain only a single copy of RSV genomic RNA, but the protein targets are often expressed at a higher copy number, each one of which can act as a potential target [22].

Here, we present the protocols for the development of a nanoparticle-amplified immuno-PCR (NPA-IPCR) assay for the detection of RSV (Fig. 1). The diagnostic assay offers a unique dual-amplification approach to achieve an exceptionally low limit of detection. The first step of signal amplification is achieved by using gold nanoparticles (AuNPs) to release multiple strands of reporter DNA per binding event. After reporter DNA is released, real-time PCR is used to provide a second level of amplification. Additionally, the assay utilizes magnetic microparticles (MMPs) to

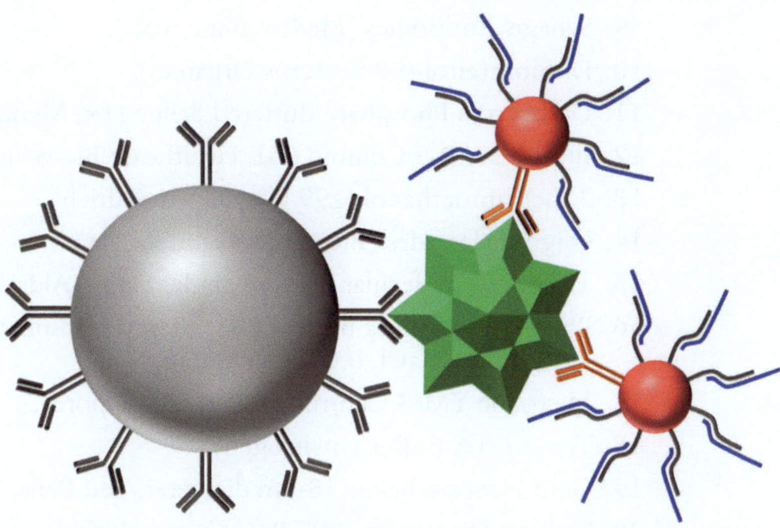

Fig. 1 Illustration of the immuno-nanoparticle-PCR sandwich. RSV antigen (*green*) is captured and separated using antibody-functionalized 1 μm magnetic microparticles (*gray*). The magnetic bead–antigen complex is then reacted with antibody–DNA-functionalized 15 nm AuNPs (*red*). Upon heating, hybridized DNA (*blue*) is released from the AuNPs and is analyzed via real-time PCR

capture antigen for improved washing to enable better target extraction. Our results suggest that the NPA-IPCR assay offers not only >1,000-fold improvement in limit of detection over enzyme-linked immunosorbent assay (ELISA) but also a fourfold improvement over detection via real-time reverse transcription PCR (real-time RT-PCR).

2 Materials

1. Nuclease-Free Molecular Grade Water (Mediatech, Inc.).
2. MagnaBind Amine Derivatized Magnetic Microparticles (Pierce Biotechnology).
3. Sulfosuccinimidyl-4-[*N*-maleimidomethyl]cyclohexane-1-carboxylate (Pierce Biotechnology).
4. Coupling buffer (CB): 50 mM phosphate buffer (pH 7.0), 0.15 M NaCl, 5 mM EDTA.
5. Analog vortex mixer (VWR).
6. MagneSphere Technology Magnetic Separation Stand (Promega).
7. Barnstead/Thermolyne Labquake Shaker/Rotisserie.
8. Anti-RSV F-protein antibodies (F-mix, clones 1269 and 1214).

9. Synagis Antibodies (MedImmune, Inc.).
10. Dithiothreitol (99 %, Acros Organics).
11. Dulbecco's Phosphate Buffered Saline (1×, Mediatech, Inc.).
12. Illustra NAP-5 Column (GE Healthcare Bio-Sciences Corp.).
13. 2-mercaptoethanol (≥99.0 %, Sigma-Aldrich).
14. Oligonucleotides (Biosearch Technologies).
15. Tween20 (molecular biology grade, Sigma-Aldrich).
16. Phosphate washing buffer (PB): 10 mM phosphate buffer (pH 7.0), 0.3 M NaCl, 0.02 % Tween20.
17. Microcon YM-3 Centrifugal Filter (Millipore).
18. Tris–EDTA Buffer (Invitrogen).
19. Gold Nanoparticles (15-nm diameter, Ted Pella, Inc.).
20. Sodium Hydroxide (≥97.0 %, Sigma-Aldrich).
21. Sodium Chloride (≥99.0 %, Sigma-Aldrich).
22. Legend Micro 21 Centrifuge (Thermo Electron Corp.).
23. HEp-2 Cells (American Type Culture Collection).
24. Respiratory syncytial virus (strain A_2).
25. OPTI-MEM Media (Gibco).
26. Fetal Bovine Serum (Atlanta Biologicals, Inc.).
27. Amphotericin-B (Sigma).
28. Gentamicin (Mediatech, Inc.).
29. L-Glutamine (Mediatech, Inc.).
30. Cell culture flasks, cell culture plates, cell scrapers, centrifuge tubes, and serological pipette tips.
31. Allegra 21R Centrifuge (Beckman Coulter).
32. Dry ice.
33. Ethanol (Pharmco-AAPER).
34. Bovine Serum Albumin (Sigma).
35. 655 nm Quantum Dots (Invitrogen).
36. BioMag 96-well Plate Side Pull Magnetic Separator (Polysciences, Inc.).
37. Axiovert 200 Inverted Fluorescence Microscope (Zeiss).
38. UV–vis Spectrophotometer (Agilent).
39. Standard Heatblock (VWR).
40. Research Quartz Crystal Microbalance (Maxtek, Inc.).
41. Rotor-Gene Q 5-Plex Thermal Cycler System (Qiagen).
42. Rotor-Gene SYBR Green PCR Master Mix (Qiagen).
43. RNeasy Mini Kit (Qiagen).

44. 1 mL syringes and 20-gauge needles.
45. RNase-Free DNase Set (Qiagen).
46. Q-Taq One-Step qRT-PCR SYBR Kit (Clontech).
47. Costar UV Microtiter Plate (Corning).
48. Goat Anti-Human HRP Conjugate Secondary Antibodies (Southern Biotech).
49. TMB One Solution (Promega).
50. Sulfuric Acid (EMD Chemicals).
51. Synergy HT Microplate Reader (BioTek).
52. RNase-free microcentrifuge tubes and pipette tips.

3 Methods

3.1 Preparation of RSV-Infected Cell Lysate

1. Supplement 0.5 L of OPTI-MEM media with 10 mL of fetal bovine serum (FBS), 5 mL of 250 μg/mL of amphotericin-B, 0.5 mL of 50 mg/mL gentamicin, and 5 mL of 200 mM L-glutamine.
2. Dilute respiratory syncytial virus (RSV) A_2 to 5 mL in supplemented media (*see* **Notes 1** and **2**).
3. Aspirate the media off of a ~90 % confluent T-150 flask of HEp-2 cells.
4. Add diluted RSV to the T-150 flask ensuring that the virus solution completely covers the bottom of the flask.
5. Incubate the T-150 flask at 37 °C with 5 % CO_2 for 1 h.
6. Add 35 mL of supplemented media to the flask.
7. Incubate the T-150 flask at 37 °C with 5 % CO_2 for 4 days.
8. Scrape infected cells from the surface of the flask.
9. Collect the supernatant containing the infected cells in a 50 mL centrifuge tube.
10. Centrifuge for 5 min at 500×*g*.
11. Remove the supernatant and resuspend the cell pellet in 5 mL of supplemented media.
12. Freeze the cells using a slurry of ethanol and dry ice in order to lyse cells and release virus particles.
13. Thaw the frozen cells in a 37 °C water bath.
14. Repeat **steps 12** and **13** two more times to ensure the release of virus particles from the cell wall.
15. Centrifuge the cell lysate at 100×*g* for 5 min to pellet large cellular debris.
16. Collect the supernatant, separate into aliquots of 0.5 mL, and store at −80 °C.

3.2 Coupling of Antibodies to Magnetic Microparticles

MagnaBind amine-derivatized magnetic microparticles (MMPs) were activated with sulfosuccinimidyl-4-[*N*-maleimidomethyl] cyclohexane-1-carboxylate (Sulfo-SMCC) to facilitate the attachment of reduced antibodies.

1. Wash 200 μL of MMPs three times by extracting MMPs with a magnetic separation stand and resuspending MMPs in 500 μL coupling buffer (CB).
2. Resuspend washed MMPs in 200 μL of CB.
3. Activate the MMPs by adding 20 μL of 1 mM Sulfo-SMCC.
4. Mix the solution and place on a rotisserie for 1 h.
5. Wash MMPs three times using the procedure described in **step 1** to remove excess Sulfo-SMCC.
6. In a separate microcentrifuge tube, combine 15 μL of 1.2 mM dithiothreitol (DTT) and 15 μL of 30 mg/mL anti-RSV F-protein antibodies (F-mix) (*see* **Note 3**).
7. Mix the solution and place on a rotisserie for 0.5 h.
8. Separate reduced antibodies from DTT using a NAP-5 column (*see* **Note 4**).
9. Combine activated MMPs with the column-purified reduced antibodies.
10. Mix the solution and place on a rotisserie for 1 h.
11. Quench the reaction by adding 2-mercaptoethanol to a final concentration of 100 μM.
12. After 1 h of quenching, wash the MMPs three times with Dulbecco's phosphate-buffered saline (PBS).
13. Resuspend washed MMPs in a final volume of 300 μL PBS.

3.3 Antibody-Magnetic Microparticle Characterization Experiment

Following attachment of antibodies to MMPs, a pull-down experiment can be performed to validate the attachment. The following protocol can be used to fluorescently label virus pulled down by functional antibodies.

1. Mix 10 μL of antibody-conjugated MMPs, 100 μL of RSV-infected cell lysate, and 100 μL of 4 % bovine serum albumin (BSA) in PBS.
2. Place on a rotisserie at room temperature for 2 h.
3. Wash the MMPs three times by extracting MMPs with a magnetic separation stand and resuspending MMPs in 500 μL of PBS.
4. Resuspend washed MMPs in 200 μL of PBS.
5. Mix MMPs with 100 μL of 2 % BSA containing 12 μg/mL of F-mix antibody with a quantum dot (QD) attached to the antibody.

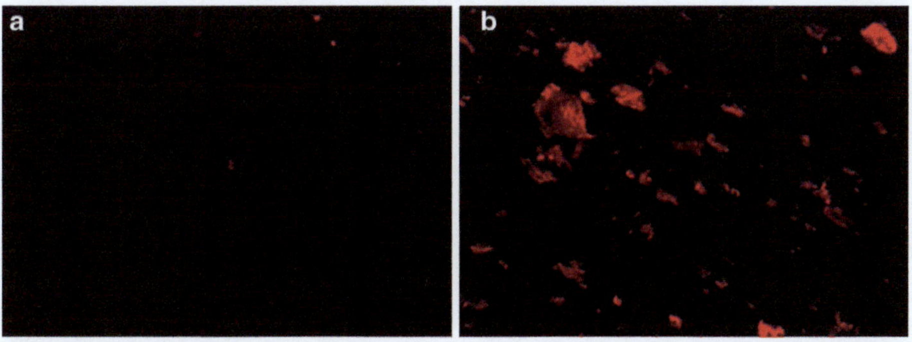

Fig. 2 Fluorescence images of MMP–virus–QD complexes. A mixture of MMPs (**a**) unconjugated or (**b**) conjugated to F-mix antibodies and 655 nm quantum dots conjugated to F-mix antibodies were mixed with RSV. Magnetic particles and associated complexes were extracted and washed. Images are 20× magnification

6. Place on a rotisserie at room temperature for 2 h.
7. Wash the MMPs using the procedure described in **step 3**.
8. Place the solution of MMP–RSV–QD complexes in a 96-well plate.
9. Place the 96-well plate on a BioMag® 96-well Plate Side Pull Magnetic Separator.
10. Image the 96-well plate on an inverted fluorescence microscope.

Figure 2 shows fluorescence images from the protocol in Subheading 3.3. Figure 2a, which contains negligible fluorescence, shows a control non-functionalized MMPs used in the protocol in Subheading 3.3. Figure 2b shows MMPs functionalized using the protocol in Subheading 3.2 and then processed using the protocol in Subheading 3.3. The fluorescence was generated by the presence of the fluorescent quantum dot bound to the MMP–RSV complexes. Formation of these complexes confirms that the antibodies remain functional after attachment to MMPs. Examining the average pixel intensity over the entire fluorescence images revealed that the functionalized MMPs (33.9 average pixel intensity) generated tenfold greater fluorescence over the control MMPs (3.2 average pixel intensity).

3.4 Coupling of Antibodies and DNA to Gold Nanoparticles

Thiolated DNA sequences were received as disulfides and were activated by cleaving the disulfide bond prior to coupling.

1. Resuspend lyophilized DNA in 100 mM DTT, 0.1 M phosphate buffer, pH 8.3.
2. Incubate the DNA at room temperature for 0.5 h to facilitate DTT reduction of disulfide bonds.
3. Desalt the reduced DNA using a Microcon YM-3 centrifugal filter (*see* **Note 5**).

Table 1
Glossary of oligonucleotide sequences

Name	Sequence 5′→3′
Comp_55	[C6Thiol] TTTTT TTTTT TTTTT GCTTG TCTCG TAAGT TGAGA TTTCG CTATG CACGG TCCTT
Tag_76	CTGCG ACGAT CTACC ATCGA CGTAC CAGGT CGGTT GAAGG ACCGT GCATA GCGAA ATCTC AACTT ACGAG ACAAG C
Tag primer 1	CTGCG ACGAT CTACC AT
Tag primer 2	GCTTG TCTCG TAAGT TGA
RSV primer 1	GCTCT TAGCA AAGTC AAGTT GAATG A
RSV primer 2	TGCTC CGTTG GATGG TGTAT T

4. Resuspend purified DNA in Tris–EDTA (TE) buffer and store in small aliquots at −80 °C.
5. Add 1 M NaOH to adjust the pH of the gold nanoparticles (AuNPs) to 9.3.
6. Add 35 μL of 0.2 mg/mL Synagis antibody diluted in water to 10 mL of 2.3 nM AuNPs (pH 9.3) (*see* **Note 6**).
7. Place AuNPs on a rotisserie at room temperature for 0.5 h.
8. Add 50.8 μL of 107 μM activated DNA (Comp_55; Table 1) to AuNPs (*see* **Note 7**).
9. Place AuNPs on a rotisserie at room temperature for 0.5 h.
10. Add 1 mL of 0.1 M phosphate buffer (pH 7.0), 2 μL of Tween20, and 200 μL 5.0 M NaCl to the AuNP solution (*see* **Note 8**).
11. Place AuNPs on a rotisserie at room temperature for 1 h.
12. Add 200 μL 5.0 M NaCl to the AuNP solution.
13. Repeat **steps 11** and **12**.
14. Place AuNPs on a rotisserie at room temperature for 1 h.
15. Remove excess DNA by centrifuging the solution for 20 min at 21,100 × *g*, and removing the supernatant.
16. Resuspend the red oily pellet of AuNPs in a stock solution of PB.
17. Repeat the washing procedure described in **steps 15** and **16** three times.
18. Resuspend the red oily pellet of AuNPs in 1 mL of PB.
19. Determine the concentration of the AuNPs via UV–vis spectroscopy (*see* **Note 9**).

20. Add sufficient Tag DNA (Tag_76; Table 1) to bring the molar ratio of Tag DNA:AuNPs to 200:1.
21. Place AuNPs on a rotisserie at room temperature overnight.
22. Remove excess Tag DNA using the washing procedure described in **steps 15–17**.
23. Resuspend the red oily pellet of AuNPs in 1 mL of PB.
24. Determine the concentration of the AuNPs via UV–vis spectroscopy (*see* **Note 9**).
25. Add PB to bring the final concentration of AuNPs to 5.0 nM.

3.5 Validation of DNA–Gold Nanoparticle Functionalization

Antibodies and DNA (Comp_55) were attached to AuNPs following the procedure stated in Subheading 3.4, stopping at **step 19** (prior to the addition of Tag DNA).

1. Add Tag DNA to 100 μL of 5 nM AuNPs at various molar ratios of Tag DNA:AuNP (1:10, 1:25, 1:50, 1:100, 1:150, 1:200, 1:300, and 1:500).
2. As a control, add the same amount of Tag DNA used in **step 1** to 100 μL of PB for each AuNP sample.
3. Place AuNPs and control samples on a rotisserie at room temperature for 24 h.
4. Centrifuge all samples for 0.5 h at 16,100 × *g* to pellet the AuNPs and Tag DNA hybridized to them.
5. Collect 90 μL of supernatant from each sample (AuNP and PB samples).
6. Determine the absorbance at 260 nm of each supernatant on a UV–vis spectrophotometer.
7. Calculate the difference between samples incubated with and without AuNPs.

Figure 3 shows quantitation of the loading of Tag DNA on AuNPs using the validation protocol in Subheading 3.5. Figure 3 revealed that saturation of the AuNPs with Tag DNA, ~70 strands per particle, occurred when Tag DNA was incubated at a molar ratio ≥400:1 of Tag DNA:AuNP. However, a much lower concentration, 200:1, still allows for the hybridization of ~50 strands per particle, which is suitable for subsequent experiments.

3.6 Quartz Crystal Microbalance Validation of Antibody–Gold Nanoparticle Attachment

All quartz crystal microbalance (QCM) experiments were performed with a flow rate of 30 μL/min and 5 MHz Ti/Au quartz crystals.

1. Allow the QCM to equilibrate for 1 h with PBS flowing over the crystal.
2. Obtain a 5 min baseline with PBS flowing over the crystal.
3. Turn off the pump to stop the flow over the crystal and change the sample inlet flow from PBS- to RSV-infected cell lysate.

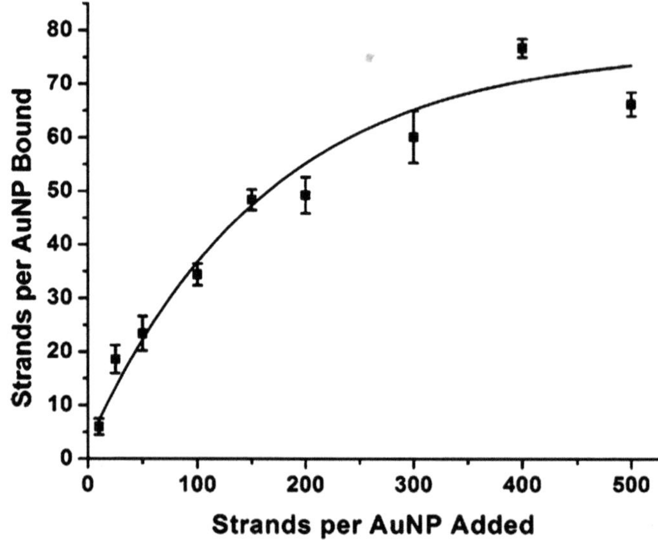

Fig. 3 Number of strands of Tag DNA bound per particle versus number of strands of Tag DNA added per particle. Data shown as the mean ± S.D. ($n = 3$)

4. Turn on the pump to resume the flow over the crystal.
5. Allow RSV-infected cell lysate to flow for 10 min while monitoring the frequency of the crystal.
6. Turn off the pump to stop the flow over the crystal and change the sample inlet flow from RSV-infected cell lysate to PBS.
7. Turn on the pump to resume the flow over the crystal.
8. Allow PBS to flow for 5 min while monitoring the frequency of the crystal.
9. Turn off the pump to stop the flow over the crystal and change the sample inlet flow from PBS to 1 % BSA in PBS.
10. Turn on the pump to resume the flow over the crystal.
11. Allow 1 % BSA to flow for 10 min while monitoring the frequency of the crystal.
12. Turn off the pump to stop the flow over the crystal and change the sample inlet flow from 1 % BSA to PBS.
13. Turn on the pump to resume the flow over the crystal.
14. Allow PBS to flow for 10 min while monitoring the frequency of the crystal.
15. Turn off the pump to stop the flow over the crystal and change the sample inlet flow from PBS to either AuNPs functionalized with anti-RSV antibodies and DNA or, as a control, AuNPs functionalized with DNA alone.
16. Turn on the pump to resume the flow over the crystal.

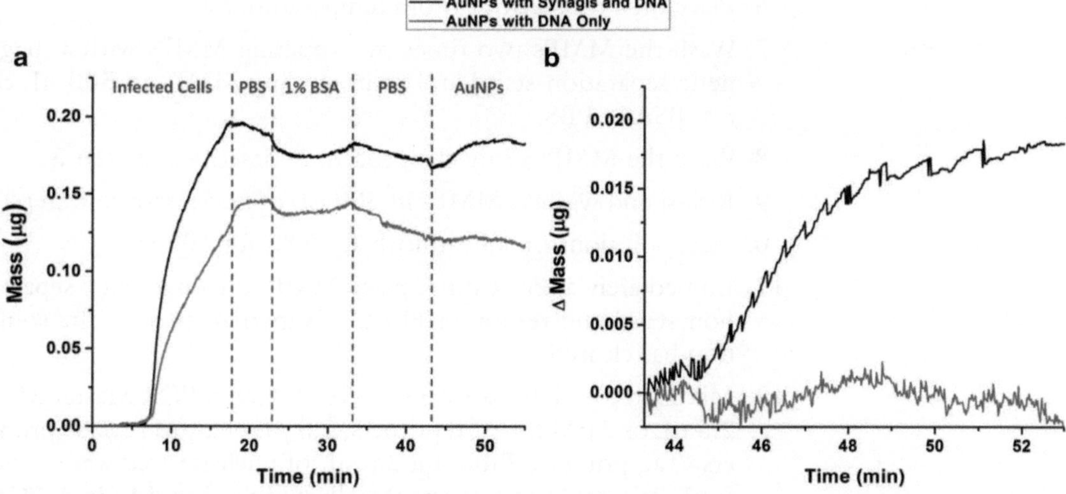

Fig. 4 Mass change of RSV exposed to AuNPs functionalized with either DNA alone (*gray*) or DNA and Synagis antibodies (*black*). (**a**) Entire experiment showing total mass bound upon flowing RSV onto the quartz crystal, followed by a PBS wash, blocking with 1 % BSA, a second PBS wash, and finally functionalized AuNPs. (**b**) The change in mass upon flowing functionalized AuNPs over the blocked and washed RSV-infected cells

17. Allow AuNPs to flow for 10 min while monitoring the frequency of the crystal.

Using the validation protocol in Subheading 3.6, it can be observed that when antibody–DNA–AuNPs are flowed along the crystal, binding occurs and can be observed by an increase in mass detected by the microbalance (Fig. 4b). However when DNA–AuNPs which are not functionalized with antibodies are flowed along the crystal, no significant change in mass is observed (Fig. 4b). By comparing the addition of AuNPs with conjugated antibodies and DNA to the addition of AuNPs with DNA only, it is apparent that not only do antibodies coupled to the AuNP facilitate binding to RSV but also that AuNPs with DNA alone have very little nonspecific binding to RSV, as evident by no change in mass under control conditions.

3.7 Nanoparticle-Amplified Immuno-PCR

1. Mix 5 μL of antibody-conjugated MMPs, 100 μL of RSV-infected cell lysate, and 200 μL of 5 % BSA in PBS.
2. Place on a rotisserie at room temperature for 1 h.
3. Wash the MMPs three times by extracting MMPs with a magnetic separation stand and resuspending MMPs in 500 μL of PBS.
4. Resuspend washed MMPs in 200 μL of PBS.
5. Mix MMPs with 5 μL of 5 nM antibody–DNA-functionalized AuNPs and 300 μL of 5 % BSA in PBS.

6. Place on a rotisserie at room temperature for 1 h.
7. Wash the MMPs two times by extracting MMPs with a magnetic separation stand and resuspending MMPs in 500 μL of 5 % BSA in PBS.
8. Wash the MMPs using the procedure described in **step 3**.
9. Resuspend washed MMPs in 300 μL of nuclease-free water.
10. Place solutions on a heatblock at 95 °C for 10 min.
11. Immediately after heating, place MMPs on a magnetic separation stand and remove 100 μL of supernatant once the solution has cleared.
12. Mix 12.5 μL of 2× Rotor-Gene SYBR Green PCR Master Mix, 2.5 μL of 2 μM forward primers, 2.5 μL of 2 μM reverse primers (Tag primers; Table 1), 2.5 μL of nuclease-free water, and 5 μL of sample (supernatant collected in **step 11**) in a PCR tube.
13. In a thermocycler, heat samples to 95 °C for 3 min to activate DNA polymerases.
14. In a thermocycler, perform 40 cycles of 95 °C for 15 s to denature, 60 °C for 60 s to anneal and extend, and 72 °C for 15 s to detect fluorescence. An example is shown in Fig. 5.
15. Obtain a melt curve by heating samples from 50 to 99 °C while detecting fluorescence.

3.8 Enzyme-Linked Immunosorbent Assay

1. Incubate 100 μL of 10 μg/mL F-mix antibodies in a well of a 96-well protein binding plate.
2. Incubate at room temperature for 1 h.
3. Gently aspirate off antibody solution from the well.
4. Turn the plate upside down and tap it on a paper towel to remove any residual solution.
5. Add 300 μL of PBS to the well.
6. Repeat the washing procedure described in **steps 3–5** two more times.
7. Repeat **steps 3** and **4**.
8. Add 300 μL of 5 % BSA to the well.
9. Incubate at room temperature for 1 h.
10. Repeat **steps 3** and **4**.
11. Add 100 μL of RSV-infected cell lysate to the well.
12. Incubate at room temperature for 1 h.
13. Perform the washing procedure described in **steps 3–5** three times.
14. Repeat **steps 3** and **4**.

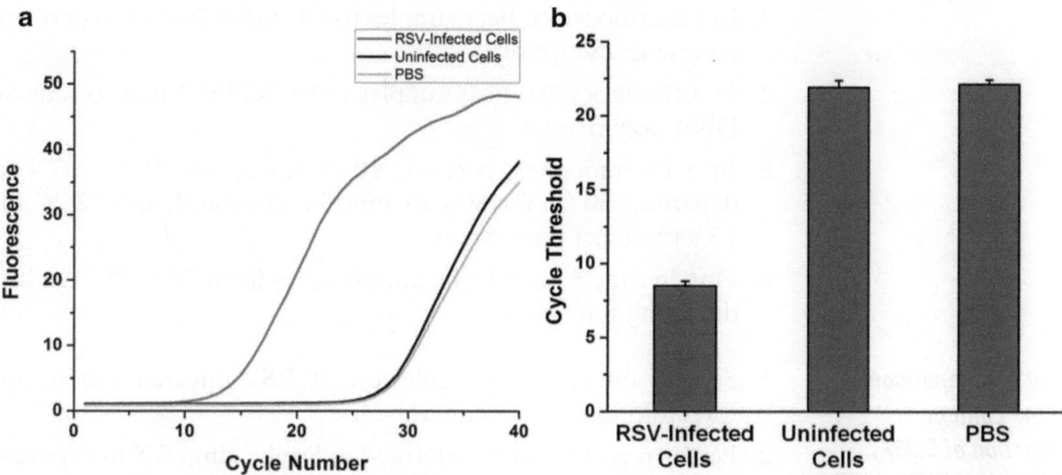

Fig. 5 RSV detection via nanoparticle-amplified immuno-PCR. (**a**) Real-time PCR results of NPA-IPCR performed on RSV-infected HEp-2 cell lysates, uninfected HEp-2 cell lysates, and PBS. (**b**) Comparison of the cycle thresholds of the same experiment. Data shown as the mean ± S.D. (*n* = 3)

15. Add 100 μL of 10 μg/mL of Synagis antibodies diluted in 5 % BSA to the well.
16. Incubate at room temperature for 1 h.
17. Perform the washing procedure described in **steps 3–5** three times.
18. Repeat **steps 3** and **4**.
19. Add 100 μL of a 1:1,000 dilution in 5 % BSA of goat anti-human HRP-conjugated secondary antibodies to the well.
20. Incubate at room temperature for 1 h.
21. Perform the washing procedure described in **steps 3–5** five times.
22. Repeat **steps 3** and **4**.
23. Add 100 μL of TMB One Solution to the well.
24. Incubate at room temperature for 10 min.
25. Add 100 μL of 2 M H_2SO_4 to the well.
26. Determine the absorbance at 450 nm of the solution on a microplate reader.

3.9 RNA Isolation and Real-Time Reverse Transcription PCR

1. Isolate RNA from RSV-infected cell lysates using an RNeasy Mini Kit (*see* **Note 10**).
2. Mix 12.5 μL of 2× One-Step qRT-PCR Buffer plus SYBR, 0.5 μL of 50× QTaq DNA Polymerase Mix, 0.4 μL of 60× qRT Mix, 2.5 μL of 2 μM forward primers, 2.5 μL of 2 μM reverse primers (RSV primers; Table 1), 1.6 μL of nuclease-free water, and 5 μL of sample (isolated RNA from **step 1**) in a PCR tube.

3. In a thermocycler, heat samples to 48 °C for 20 min to perform reverse transcription.
4. In a thermocycler, heat samples to 95 °C for 3 min to activate DNA polymerases.
5. In a thermocycler, perform 40 cycles of 95 °C for 15 s to denature, 60 °C for 60 s to anneal and extend, and 72 °C for 15 s to detect fluorescence.
6. Obtain a melt curve by heating samples from 50 to 99 °C while detecting fluorescence.

3.10 Comparison of the Limit of Detection of ELISA, RT-PCR, and NPA-IPCR

1. Serially dilute a stock solution of RSV-infected cell lysates threefold in supplemented media.
2. Perform an ELISA as described in Subheading 3.8 in triplicate on serially diluted (threefold) RSV-infected cell lysates.
3. Perform an ELISA as described in Subheading 3.8 in triplicate on uninfected HEp-2 cell lysates.
4. Using the original stock solution used in **step 1**, serially dilute RSV-infected cell lysates tenfold in supplemented media.
5. Perform RT-PCR as described in Subheading 3.9 in triplicate on serially diluted (tenfold) RSV-infected cell lysates.
6. Perform RT-PCR as described in Subheading 3.9 in triplicate on uninfected HEp-2 cell lysates.
7. Perform NPA-IPCR as described in Subheading 3.7 in triplicate on serially diluted (tenfold) RSV-infected cell lysates.
8. Perform NPA-IPCR as described in Subheading 3.7 in triplicate on uninfected HEp-2 cell lysates.
9. Determine the limit of detection of each assay by calculating three standard deviations above the background (uninfected HEp-2 cell lysates).

The normalized signal is obtained by subtracting the background and dividing each value by the response obtained at the highest virus concentration. Both NPA-IPCR and RT-PCR have a substantially lower limit of detection than ELISA (Fig. 6). The lower limit of detection for ELISA is 16,000 PFU/mL, while RT-PCR is 17.9 PFU/mL and NPA-IPCR is 4.1 PFU/mL. Although the dilution curves for NPA-IPCR and RT-PCR appear similar, the lower dilutions, shown in Fig. 6b, provide additional insight. At lower virus concentrations, ≤8.3 PFU/mL, RT-PCR has more variation between replicates. While both assays give an observable shift at 8.3 PFU/mL, only the NPA-IPCR is statistically above the limit of detection. The variability between replicates is due in part to the amount of amplification that is required to reach the cycle threshold but may also be due to other properties of PCR [23]. At a concentration of 8.3 PFU/mL, RT-PCR requires

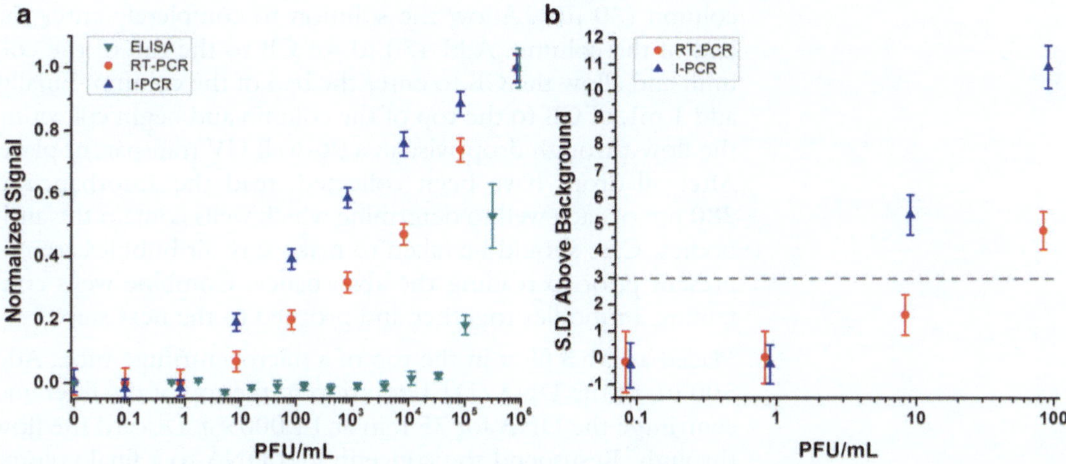

Fig. 6 Detection of decreasing concentrations of RSV. (**a**) Comparison of RSV detection using ELISA, real-time RT-PCR, and the developed NPA-IPCR assay. (**b**) The number of standard deviations above background using both RT-PCR and NPA-IPCR at low virus concentrations. The *gray dashed line* shows a 3σ limit of detection. Data shown as the mean ± S.D. ($n = 3$). NPA-IPCR data has been shifted to prevent overlapping with RT-PCR data

29.1 cycles to reach the threshold while NPA-IPCR requires only 19.4 cycles. NPA-IPCR effectively moved the low concentration dilutions up 10 cycles, in turn reducing the amount of PCR-induced variation between replicates.

4 Notes

1. Respiratory syncytial virus is a human pathogen and requires handling in a biosafety level 2 (BSL-2) facility. Use necessary precautions.
2. Virus should be stored at −80 °C and rapidly thawed in a 37 °C water bath immediately prior to use. Once the viral titer is determined, additional freeze/thaw cycles should be avoided as they can result in a change in the number of PFUs. Therefore, it is best to store viral standards in small (≤1.0 mL) aliquots so thawed virus that is unused can be appropriately discarded.
3. Antibody reduction should be coordinated so that once the MMPs are activated, they can be immediately used.
4. Care should be taken to make sure the column does not dry out at any time during the process. Allow excess storage buffer to flow through the NAP-5 column. Add 2.5 mL of CB to the top of the column and allow to flow through the column. Continue adding CB and allow it to equilibrate the column until a total of 10 mL of CB has passed through the column. Allow the residual CB to completely enter the bed of the column and then add the antibody solution to the top of the

column (30 μL). Allow the solution to completely enter the bed of the column. Add 470 μL of CB to the top of the column and allow the CB to enter the bed of the column. Finally, add 1 mL of CB to the top of the column and begin collecting the flow-through drop-wise in a 96-well UV transparent plate. After all drops have been collected, read the absorbance at 280 nm of each well to determine which wells contain the antibodies. Care should be taken to make sure air bubbles are not present prior to reading the absorbance. Combine wells containing antibodies together and proceed to the next step.

5. Placed a YM-3 filter in the top of a microcentrifuge tube. Add 500 μL of the DNA/DTT solution to the top of the filter and centrifuge the DNA for 75 min at 14,000 × *g*. Discard the flow through. Resuspend the concentrated DNA to a final volume of 500 μL in 1× TE buffer and centrifuge a second time. Once again, resuspend the concentrated DNA to a final volume of 500 μL in 1× TE buffer and centrifuge. Place the filter upside down in a clean microcentrifuge tube and centrifuge for 3 min at 1,000 × *g*. Remove the filter and dilute the DNA to the desired concentration in TE buffer and store at −80 °C in small aliquots prior to use.
6. 15 nm AuNPs are shipped at approximately 2.3 nM. Prior to coupling, the concentration should be verified using UV–vis spectroscopy and a molar extinction coefficient at 520 nm of 3.64×10^{8} M^{-1} cm^{-1}.
7. Empirical evidence has revealed that a minimum of 200 strands of DNA per 15 nm AuNP is necessary for adequate coverage of the AuNP. At lower concentrations, coverage may be incomplete and result in aggregation of AuNPs. If the DNA concentration used differs from the concentration in Subheading 3.4, **step 8**, the volume added should be adjusted to bring the final DNA:AuNP molar ratio to 200:1.
8. Slowly increasing the salt concentration is essential for adequate loading of the AuNP. In Subheading 3.4, **step 10**, NaCl should only be added after phosphate buffer and Tween20 have been added. If the solution turns from red to a shade of purple, aggregation has occurred. Adding NaCl drop-wise with vortexing between drops can help reduce aggregation.
9. Prior to determining the concentration of AuNPs, brief sonication can be used to resuspend AuNPs that pellet during centrifugation. Any pellet that does not resuspend upon sonication has most likely irreversible aggregated.
10. In a typical RNA isolation, 300 μL of cell lysate was mixed with 400 μL lysis buffer (RLT) and homogenized by passing the sample through a 20 gauge needle eight times. 700 μL of 70 % was then added to the sample, and it was bound to an

RNeasy spin column. On the column, DNA was digested using an RNase-Free DNase Set. The remaining RNA was then washed with the appropriate buffers (RW1 and RPE). After washing, the RNA was eluted in 50 μL of nuclease-free water and stored in small aliquots at −80 °C. Purified RNA samples should be stored in small enough aliquots to avoid multiple freeze/thaw cycles as this may lead to degradation.

References

1. Robert CW (2003) Review of epidemiology and clinical risk factors for severe respiratory syncytial virus (RSV) infection. J Pediatr 143:112–117
2. Shay DK, Holman RC, Newman RD, Liu LL, Stout JW, Anderson LJ (1999) Bronchiolitis-associated hospitalizations among US children, 1980–1996. J Am Med Assoc 282:1440–1446
3. Falsey AR, Walsh EE (2005) Respiratory syncytial virus infection in elderly adults. Drugs Aging 22:577–587
4. Falsey AR, Hennessey PA, Formica MA, Cox C, Walsh EE (2005) Respiratory syncytial virus infection in elderly and high-risk adults. N Engl J Med 352:1749–1759
5. Hall CB (2001) Respiratory syncytial virus and parainfluenza virus. N Engl J Med 344:1917–1928
6. Moore E, Barber J, Tripp R (2008) Respiratory syncytial virus (RSV) attachment and nonstructural proteins modify the type I interferon response associated with suppressor of cytokine signaling (SOCS) proteins and IFN-stimulated gene-15 (ISG15). Virol J 5:116
7. Tripp RA (2004) Pathogenesis of respiratory syncytial virus infection. Viral Immunol 17:165–181
8. Chin J, Magoffin RL, Shearer LA, Schieble JH, Lennette EH (1969) Field evaluation of a respiratory syncytial virus vaccine and a trivalent parainfluenza virus vaccine in a pediatric population. Am J Epidemiol 89:449–463
9. Kapikian AZ, Mitchell RH, Chanock RM, Shvedoff RA, Stewart CE (1969) An epidemiologic study of altered clinical reactivity to respiratory syncytial (RS) virus infection in children previously vaccinated with an inactivated RS virus vaccine. Am J Epidemiol 89:405–421
10. Kim HW, Canchola JG, Brandt CD, Pyles G, Chanock RM, Jensen K, Parrott RH (1969) Respiratory syncytial virus disease in infants despite prior administration of antigenic inactivated vaccine. Am J Epidemiol 89:422–434
11. Kaur J, Tang RS, Spaete RR, Schickli JH (2008) Optimization of plasmid-only rescue of highly attenuated and temperature-sensitive respiratory syncytial virus (RSV) vaccine candidates for human trials. J Virol Methods 153:196–202
12. Committee on Infectious Diseases, Committee on Fetus and Newborn (1998) Prevention of respiratory syncytial virus infections: indications for the use of palivizumab and update on the use of RSV-IGIV. Pediatrics 102:1211–1216
13. Meissner HC, Bocchini JA Jr, Brady MT, Hall CB, Kimberlin DW, Pickering LK (2009) The role of immunoprophylaxis in the reduction of disease attributable to respiratory syncytial virus. Pediatrics 124:1676–1679
14. Committee on Infectious Diseases (1996) Reassessment of the indications for ribavirin therapy in respiratory syncytial virus infections. Pediatrics 97:137–140
15. Goodrich JS, Miller MB (2007) Comparison of Cepheid's analyte-specific reagents with BD directigen for detection of respiratory syncytial virus. J Clin Microbiol 45:604–606
16. Perkins SM, Webb DL, Torrance SA, El Saleeby C, Harrison LM, Aitken JA, Patel A, DeVincenzo JP (2005) Comparison of a real-time reverse transcriptase PCR assay and a culture technique for quantitative assessment of viral load in children naturally infected with respiratory syncytial virus. J Clin Microbiol 43:2356–2362
17. Chomczynski P (1992) Solubilization in formamide protects RNA from degradation. Nucleic Acids Res 20:3791–3792
18. Hargrove JL, Schmidt FH (1989) The role of mRNA and protein stability in gene expression. FASEB J 3:2360–2370
19. Mahony JB (2008) Detection of respiratory viruses by molecular methods. Clin Microbiol Rev 21:716–747
20. Bentzen EL, House F, Utley TJ, Crowe JE, Wright DW (2005) Progression of respiratory syncytial virus infection monitored by fluorescent quantum dot probes. Nano Lett 5:591–595
21. Adler M, Schulz S, Fischer R, Niemeyer CM (2005) Detection of Rotavirus from stool

samples using a standardized immuno-PCR ("Imperacer") method with end-point and real-time detection. Biochem Biophys Res Commun 333:1289–1294

22. Barletta J, Bartolome A (2007) Immuno-polymerase chain reaction as a unique molecular tool for detection of infectious agents. Expert Opin Med Diagn 1:267–288

23. Stowers CC, Haselton FR, Boczko EM (2010) An analysis of quantitative PCR reliability through replicates using the Ct method. J Biomed Sci Eng 3:459–469

Chapter 9

Gold Nanoparticle–Oligonucleotide Conjugates for the Profiling of Malignant Melanoma Phenotypes

John W. Stone, Reese Harry, Owen Hendley, and David W. Wright

Abstract

This chapter discusses the preparation and subsequent profiling capabilities of gold nanoparticle-oligonucleotide conjugates for multiple melanoma mRNA targets. We will outline the attachment of DNA hairpins modified with a thiol for facile attachment to gold nanoparticle surfaces through gold-sulfur bond formation. Furthermore, the ability of these conjugates to detect and distinguish phenotypic variations utilizing several melanoma cell lines and the nonmalignant cell line, HEp-2, will be investigated using flow cytometry and RT-PCR analytical techniques. The behavior of the housekeeping probe β-actin will also be investigated as a control.

Key words Gold nanoparticles, Hairpin oligonucleotides, Melanoma profiling, RT-PCR, Flow cytometry

1 Introduction

Melanoma is one of the fastest diagnosed cancers with over 70,000 new cases this year alone [1]. Early diagnosis has proven key in long-term survival rates with an 80 % survival rate in patients diagnosed in early stages of the disease [2]. Current detection methods include histological examination, reverse transcription-polymerase chain reaction (RT-PCR), and antibody-assisted protein detection [3–6]. However, these assays suffer from inconsistencies and tedious sample preparation and processing [6, 7]. More recently, live cell-imaging probes including linear oligonucleotides, hairpin oligonucleotides, and Förster resonance energy reporters have been investigated as alternative detection methods [8–11]. While potentially effective, these probes require harsh transfection reagents for cell internalization and free oligonucleotide probes are susceptible to nuclease degradation [8, 12–16].

As a consequence of the above drawbacks to current detection methods, recent efforts have focused on nanotechnology-driven approaches toward probe design. By coupling

Sandra J. Rosenthal and David W. Wright (eds.), *NanoBiotechnology Protocols*, Methods in Molecular Biology, vol. 1026, DOI 10.1007/978-1-62703-468-5_9,

oligonucleotides to the surface of gold nanoparticles, many of the current probe flaws may be attenuated or eliminated [17, 18]. For example, gold nanoparticle-oligonucleotide conjugates do not require transfection reagents for cell entry and nuclease degradation is reduced resulting from steric hindrances of the gold nanoparticles [17, 18].

Tyrosinase (TYR) is the enzyme responsible for the conversion of tyrosine to melanin as part of the melanin biosynthesis process [19]. Since TYR expression is largely specific to melanocytes and melanoma cells, it is an attractive candidate for the detection of metastatic melanoma present in blood, lymph nodes, or other organs. Additionally, the serological marker, melanoma inhibitory activity protein (MIA), and S100β, a protein which binds calcium in brain astrocytes and malignant melanoma, were conjugated to the surface of gold nanoparticles to highlight the versatility of these composites for profiling multiple mRNA targets. This ability is critical since multiple gene mutations will occur potentially initiating metastasis that could alter melanoma progression. In this chapter we will outline protocols for the preparation, characterization, and utility of gold nanoparticle-oligonucleotide probes for the detection of three different melanoma expression markers resulting from phenotypic variations between four different melanoma cell lines and the non-melanocytic HEp-2 cell line.

2 Materials

1. SK-MEL-28 cells (American Type Culture Collection, ATCC).
2. HEp-2 cells (American Type Culture Collection, ATCC).
3. Wm115 cells (provided by Vanderbilt Ingram Cancer Center).
4. MeWo cells (provided by Vanderbilt Ingram Cancer Center).
5. HS-294T cells (provided by Vanderbilt Ingram Cancer Center).
6. 15 nm gold colloid solutions (Ted Pella, Inc., Redding, CA).
7. Syto-13 nuclear dye (Invitrogen Corp, Carlsbad, CA).
8. Sytox Green viability dye (Invitrogen Corp, Carlsbad, CA).
9. Mito-tracker Orange (Invitrogen Corp, Carlsbad, CA).
10. Sterile nuclease free phosphate buffered saline without $MgCl_2$ and $CaCl_2$ (pH 7.4) (Invitrogen Corp, Carlsbad, CA).
11. Bovine deoxyribonuclease I (Promega, Madison, WI).
12. Molecular beacons (Biosearch Technologies, Inc, Novato, CA).

13. Unmodified oligonucleotides (Biosearch Technologies, Inc, Novato, CA). Custom oligonucleotides (Biosearch Technologies, Inc, Novato, CA).
14. PCR primers (Biosearch Technologies, Inc, Novato, CA).
15. 0.1 M dithiothreitol (DTT, Sigma Aldridge, St. Louis, MO).
16. 1× Tris–EDTA (TE) buffer (pH 8) (Sigma Aldridge, St. Louis, MO).
17. Tween-20 (Sigma Aldridge, St. Louis, MO).
18. 10 mM phosphate buffer (pH 7) (Sigma Aldridge, St. Louis, MO).
19. 5 M NaCl (Sigma Aldridge, St. Louis, MO).
20. 3K molecular weight centrifugation filters (Amicon Ultracel 3K, Millipore).
21. High glucose DMEM growth medium (10 % fetal bovine serum, 100 units penicillin/streptomycin, 2 % l-glutamine).
22. RPMI 1540 growth solution (10% fetal bovine serum, 100 units penicillin/streptomycin, 2 % l-glutamine).
23. Phenol red-free RPMI 1640 media (5 % fetal bovine serum).
24. 35 mm Mattek microwell dish or well plates.
25. 0.05 % trypsin–EDTA.
26. RNeasy Mini Kit (Qiagen).
27. Nuclease reaction buffer (100 mM Tris–HCl, 5 mM $MgSO_4$, 1 mM $CaCl_2$, pH 8).
28. Malvern Zetasizer Nano ZS.
29. Absorption and fluorescence plate reader.
30. UV-vis spectrometer.
31. Custom Becton Dickinson five-laser LSRII analytical flow cytometer.
32. Rotor-Gene Q real time cycler (Qiagen) and QTaq One-Step qRT-PCR SYBR Kit (Clontech).
33. 1.2 % agarose E-gel (Invitrogen).

3 Methods

3.1 Preparation of Hairpin Oligonucleotides

Purchased oligonucleotides maintain an unreduced 5′ thiol moiety which is reduced prior to addition to gold nanoparticle solutions. This step is required in order to provide available thiols for gold-sulfur bond formation.

1. 0.1 M DTT was added to the purchased hairpin DNA resulting in a 100 μM oligonucleotide solution and mixed for 30 min in order to facilitate reduction at the 5′ thiol end.

2. Reducing agent was removed via centrifugation through a 3K molecular weight centrifugal filter and subsequently washed three times with PBS.
3. Reduced oligonucleotides were resuspended in 1× Tris–EDTA (TE) buffer (pH 8) and stored at −80 °C until ready for use.

3.2 Synthesis of Gold Nanoparticle-Oligonucleotide (hAuNP) Composites

Oligonucleotide hairpins were conjugated to 15 nm gold nanoparticles through gold–sulfur bond formation via displacement of citrate. The addition of phosphate buffer, Tween-20, and increasing amounts of NaCl further stabilize the conjugates allowing for a higher number of hairpins per particle. Once prepared gold hAuNP conjugates were stored in PBS at 4 °C.

3.2.1 Conjugation of Hairpin Oligonucleotides (S100β, MIA, β-Actin) to Gold Nanoparticles

1. To 1 mL of a 10 nM citrate-stabilized gold nanoparticle solution was added DNA at a final DNA concentration of 2 μM. The solution was protected from light and mixed overnight at room temperature.
2. Following overnight incubation, 110 μL 10 mM phosphate buffer (pH 7), 1.1 μL Tween-20 (0.1 %), and 22.2 μL 5 M NaCl were added, mixed, and incubated for ≥2 h.
3. Following a minimum 2 h incubation, 22.4 μL 5 M NaCl was added, mixed, and incubated for ≥2 h.
4. Following a minimum 2 h incubation, 22.6 μL 5 M NaCl was added, mixed, and incubated for ≥2 h.
5. The resulting gold nanoparticle composites were purified via three rounds of centrifugation with washing in PBS (30 min, 17,000 × *g*) and finally stored in PBS (pH 7.4) at 4 °C.
6. Concentrations of gold nanoparticle-oligonucleotide conjugates were determined by measuring the absorption at 520 nM with an extinction coefficient of 3.64×10^{8} M^{-1} cm^{-1}.

3.3 Characterization of Gold Nanoparticle-Oligonucleotide Composites

In order to quantitate the number of hairpins per particle, DNA was cleaved from the nanoparticle surface via disulfide reduction using DTT. The fluorescence of cleaved DNA was measured on a plate reader and compared to a standard curve of DNA at known concentrations. Knowing both the nanoparticle and DNA concentrations, an average number of hairpins were estimated based on a molar ratio. Furthermore, fluorescence measurements were carried out to assay the selectivity of the hairpin probes by monitoring signal to noise values of the probe in the presence of target complement and target mismatch (MM), respectively.

3.3.1 Quantitation of Number of Hairpins per Gold Nanoparticle

1. A reducing buffer for DNA cleavage was prepared by adding 38 mL 0.1 M dibasic phosphate buffer stock together with 2 mL 0.1 M monobasic phosphate buffer resulting in a final buffer pH of 8.3. 0.617 g DTT was dissolved into 40 mL of the prepared phosphate buffer.

2. 150 μL of a 10 nM hAuNP in PBS was centrifuged at 17,000 × *g* for 10 min and resuspended in 30 μL DNase-free water.
3. The above sample is then aliquoted into three tubes each containing 10 μL followed by a tenfold dilution into the above prepared phosphate buffer.
4. The samples were mixed for 1 h followed by centrifugation at 21,000 × *g* for 10 min.
5. The supernatant containing the hairpin DNA is analyzed using a fluorescence plate reader, and absorption values were measured and compared against a calibration curve of known DNA concentrations.

3.3.2 Gold Nanoparticle-Oligonucleotide Target Selectivity

1. To a 1 nM (PBS, 0.1 % Tween-20) solution of hAuNPs was added target complement or target MM respectively to a final target concentration of 1 μM.
2. The above solutions were sealed, protected from light, and allowed to incubate for 2 h at 37 °C.
3. Signal to noise was measured by taking the ratio of target fluorescence and MM background fluorescence.

3.4 Flow Cytometry Measurements

In order to assess the ability of the different probes to discriminate varying phenotypes of malignant melanoma, three different probes—TYR, MIA, and S100β—and the housekeeping probe β-actin were used in combination with the four cell lines, SK-MEL-28, Wm115, MeWo, and HS-294T, and the non-melanocytic HEp-2 cell line. Figure 1 represents hAuNP expression values across the multiple cell lines, indicating that TYR and MIA expression values are higher in all malignant cell lines as compared to HEp-2 cells with TYP expression especially high in the SK-MEL-28 cell line. S100β, conversely, was found to possess higher expression values in the nonmalignant HEp-2 cell line, a finding not supported in the literature. Table 1 is a quantitative summary of the flow cytometry results.

1. Prepare hAuNPs (*see* Subheading 3.2.1).
2. All cell lines were plated in 6-well plates (in complete media) such that a cell concentration of 1e6 cells/well was achieved (~50 % confluence).
3. Cells were dosed with 0.25 nM melanoma hAuNP targets and 0.25 nM β-actin hAuNPs simultaneously and allowed to incubate for 4 h at 37 °C and 5 % CO_2.
4. Following the initial incubation, cells were washed twice with PBS and fresh media was added to each well and incubated for an additional 12 h at 37 °C and 5 % CO_2.

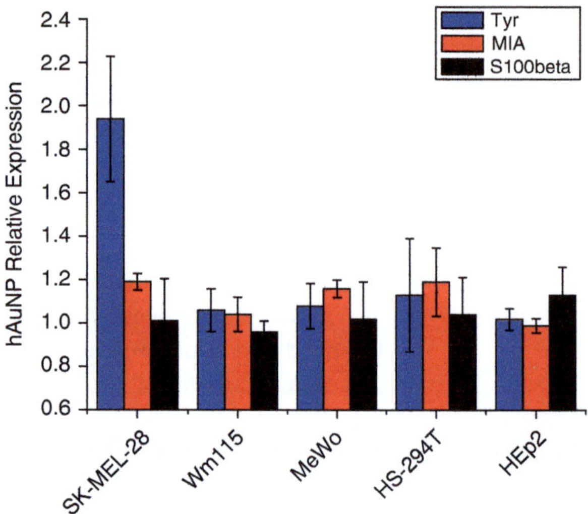

Fig. 1 hAuNP expression values for multiple melanoma cell lines. (*Blue*) tyrosinase hAuNP expression values, (*red*) MIA hAuNP expression values, and (*black*) S100 hAuNP expression values

Table 1
hAuNP expression values for multiple melanoma cell lines

hAuNP melanoma expression values			
	TYR	MIA	S100β
SK-MEL-28	1.94 ± 0.29	1.19 ± 0.04	1.01 ± 0.19
Wm115	1.06 ± 0.10	1.04 ± 0.08	0.96 ± 0.05
MeWo	1.08 ± 0.10	1.16 ± 0.04	1.02 ± 0.17
HS-294T	1.13 ± 0.26	1.19 ± 0.16	1.04 ± 0.17
HEp-2	1.02 ± 0.05	0.99 ± 0.03	1.13 ± 0.13

5. Following the 12 h incubation, cells were lifted from the plate with 0.05 % trypsin–EDTA, washed twice with cold PBS, resuspended in 500 μL cold phenol red-free media containing 5 % FBS, and stored on ice until analysis.
6. Sytox Green was added to each well at a concentration of 50 nM in order to access cell viability in the presence of hAuNPs.
7. Samples were analyzed via flow cytometry measurements with forward and side scatter. Both Sytox Green and CY five emissions (associated with hAuNPs) were collected and mean fluorescence values recorded.

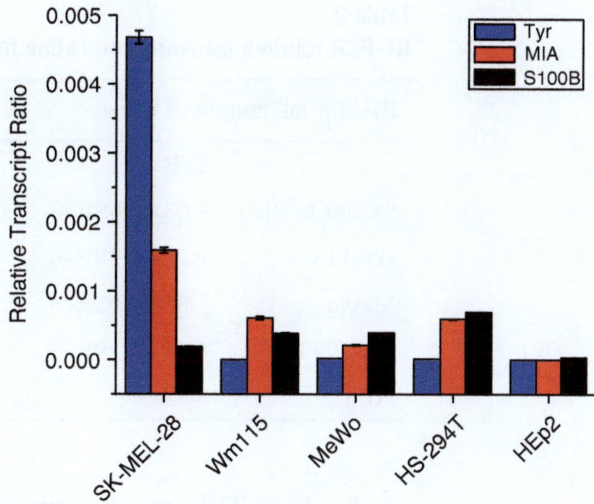

Fig. 2 RT-PCR relative transcription ratios for experimental cell lines

3.5 Reverse Transcription-Polymerase Chain Reaction (RT-PCR)

RT-PCR measurements were performed on all cell lines in order to assay relative expression levels for each of the melanoma markers in an attempt to validate hAuNPs as promising probes for melanoma mRNA detection. RT-PCR is a common technique for assaying the presence of mRNA for a given system and so provides a reasonable standard for comparison purposes. Data for TYR expression in SK-MEL-28 cells correlated well with the hAuNP probes having high expression values (Fig. 2). Neither MIA nor S100β was found to significantly discriminate between various phenotypes in the remaining malignant cell lines although their expression values were larger than those measured for HEp-2 cells. Table 2 is a quantitative representation of the cycle threshold (Ct) expression values. It is possible that the overall lack of expression in malignant cells, other than SK-MEL-28 cells, may be influenced by the individual cell types assayed and requires further investigation.

1. Cells were plated in a 6-well plate (*see* Subheading 3.4).
2. Cellular mRNA was extracted from each of the cell lines using RNeasy Mini Kit following the manufacture's protocol (1e6 cells).
3. 15 ng (5 μL) of RNA was added to a 25 μL volume with the following reagents: 12.5 μL of 2× One-Step qRT-PCR Buffer containing SYBR, 0.5 μL 50× QTaq DNA Polymerase Mix, 0.4 μL 60× qRT Mix, 200 nM forward and reverse primers (*see* **Note 1**), and DNase/RNase-free water.
4. Samples were cycled as follows: 45 min at 48 °C, 3 min at 95 °C, followed by 40 1 min cycles at 55 °C, and a final cycle for 2 min at 72 °C.

Table 2
RT-PCR relative transcription ratios for multiple melanoma cell lines

RT-PCR melanoma Ct values			
	TYR	MIA	S100β
SK-MEL-28	4.68 ± 0.09E−3	1.59 ± 0.04E−3	1.96 ± 0.06E−4
Wm115	1.26 ± 0.06E−6	6.10 ± 0.23E−4	3.86 ± 0.06E−4
MeWo	1.85 ± 0.01E−5	2.17 ± 0.05E−4	3.94 ± 0.04E−4
HS-294T	7.74 ± 0.20E−6	5.93 ± 0.06E−4	6.95 ± 0.02E−4
HEp-2	N/A	N/A	3.89 ± 0.03E−5

5. 5 μL of TYR and β-actin PCR products were subsequently assayed via electrophoresis on a 1.2 % agarose E-gel and imaged under UV light.
6. Relative transcript values were reported based on the equation $2^{-\Delta Ct}$ where ΔCt is equal to Ct target gene-Ct β-actin gene.

4 Note

1. Forward and reverse primers utilized (*see* Subheading 3.5) are as follows: TYR forward (5′-ttggcagattgtctgtagcc-3′), TYR reverse (5′-aggcattgtgcatgctgctt-3′), S100β forward (5′-atgtctgagctggagaaggccat-3′), S100β reverse (5′-actgcctgccacgagttctttgaa-3′), MIA forward (5′-catgcatgcggtcctatgcccaagctg-3′), MIA reverse (5′-gataagctttcactggcagtagaaatc-3′), β-actin forward (5′-gcgggaaatcgtgcgtgacatt-3′) and β-actin reverse (5′-gatggagttgaaggtagtttcgtg-3′).

References

1. National Cancer Institute (2006) National Foundation For Cancer Research. Research for a Cure (https://www.nfcr.org/skin-cancer)
2. Goetz T (2009) The Truth About Cancer. WIRED Magazine Issue 17.01
3. Andres R, Mayordomo JI, Visus C, Isla D, Godino J, Escudero P, Saenz A, Ortega E, Lastra R, Lambea J, Aguirre E, Elosegui L, Marcos I, Ruiz-Echarri M, Millastre E, Saez-Gutierrez B, Asin L, Vidal MJ, Ferrer A, Giner A, Larrad L, Carapeto FJ, Tres A (2008) Prognostic significance and diagnostic value of protein S-100 and tyrosinase in patients with malignant melanoma. Am J Clin Oncol 31:335–339
4. Gry M, Rimini R, Stromberg S, Asplund A, Ponten F, Uhlen M, Nilsson P (2009) Correlations between RNA and protein expression profiles in 23 human cell lines. BMC Genomics 10:365
5. Mocellin S, Rossi CR, Pilati P, Nitti D, Marincola FM (2003) DNA array-based gene profiling. Trends Mol Med 9:189–195
6. Nezos A, Lembessis P, Sourla A, Pissimissis N, Gogas H, Koutsilieris M (2009) Molecular markers detecting circulating melanoma cells by reverse transcription polymerase chain reaction: methodological pitfalls and clinical relevance. Clin Chem Lab Med 47:1–11
7. de Vries TJ, Fourkour A, Punt CJ, Diepstra H, Ruiter DJ, van Muijen GN (1999) Melanoma-inhibiting activity (MIA) mRNA is not exclusively transcribed in melanoma cells: low levels

of MIA mRNA are present in various cell types and in peripheral blood. Br J Cancer 81: 1066–1070

8. Bao G, Rhee WJ, Tsourkas A (2009) Fluorescent probes for live-cell RNA detection. Annu Rev Biomed Eng 11:25–47
9. Santangelo P, Nitin N, Bao G (2006) Nanostructured probes for RNA detection in living cells. Ann Biomed Eng 34:39–50
10. Santangelo P, Nitin N, LaConte L, Woolums A, Bao G (2006) Fluorescent probes for live-cell RNA detection. J Virol 80:682–688
11. Santangelo PJ, Nitin N, Bao G (2005) Direct visualization of mRNA colocalization with mitochondria in living cells using molecular beacons. J Biomed Opt 10:44025
12. Chen AK, Behike MA, Tsourkas A (2007) Avoiding false-positive signals with nuclease-vulnerable molecular beacons in single living cells. Nucleic Acids Res 35:1–12
13. Walev I, Bhakdi SC, Hofmann F, Djonder N, Valeva A, Aktories K, Bhakdi S (2001) Proc Natl Acad Sci USA 98:3185–3190
14. Tsourkas A, Behlke MA, Bao G (2003) Hybridization kinetics and thermodynamics of molecular beacons. Nucleic Acids Res 31:5168–5174
15. Tsourkas A, Behlke MA, Rose SD, Bao G (2003) Hybridization kinetics and thermodynamics of molecular beacons. Nucleic Acids Res 31:1319–1330
16. Tyagi S, Kramer FR (1996) Molecular beacons: probes that fluoresce upon hybridization. Nat Biotechnol 14:303–308
17. Rosi NL, Giljohann DA, Thaxton CS, Lytton-Jean AKR, Han MS, Mirkin CA (2006) Oligonucleotide-modified gold nanoparticles for intracellular gene regulation. Science 312:1027–1030
18. Seferos DS, Giljohann DA, Hill HD, Prigodich AE, Mirkin CA (2007) Gold nanoparticles delivery in mammalian live cells: a critical review. J Am Chem Soc 129:15477–15479
19. Hearing VJ, Tsukamoto K (1991) Enzymatic control of pigmentation in mammals. FASEB J 5:2902–2909

Chapter 10

Methods for Isolating RNA Sequences Capable of Binding to or Mediating the Formation of Inorganic Materials

Carly Jo Carter, Alina Owczarek, and Daniel L. Feldheim

Abstract

The ability of oligonucleotides to mediate the formation of inorganic materials is now well established. RNA and DNA are proving to be capable of mediating the formation of inorganic nanoparticles with sequence-specific control over nanoparticle size, shape, and even atomic-level crystallinity. Here we describe methods for isolating specific RNA sequences from large random sequence libraries that either bind to precut inorganic crystal wafers with high affinity or influence inorganic crystal growth.

Key words RNA, Biomineralization, Nanoparticles, SELEX

1 Introduction

Biological systems are capable of synthesizing a remarkably diverse range of inorganic materials within a relatively narrow set of reaction conditions: neutral pH, aqueous solutions, and near-ambient temperatures and pressures. Biological systems have adapted within these constraints by applying a large and dynamic combinatorial approach to materials synthesis, resulting in protein enzymes that catalyze inorganic reactions and direct the atomic-level structures and macroscopic morphologies of growing crystals. In addition to affording an organism structural integrity and protection against predation, many materials found in nature have more sophisticated physical properties such as magnetism and light focusing [1, 2].

The biomaterials found in nature have inspired a growing research effort aimed at understanding the mechanisms by which these materials are assembled, and exploiting biomolecules in the synthesis of novel materials and assembly of functional devices. In an attempt to mimic natural evolutionary processes, in vitro selection methods employing large random sequence RNA, DNA, peptide, and even whole cell libraries are now being used to discover biomolecules that

Sandra J. Rosenthal and David W. Wright (eds.), *NanoBiotechnology Protocols*, Methods in Molecular Biology, vol. 1026, DOI 10.1007/978-1-62703-468-5_10, © Springer Science+Business Media New York 2013

1. 5'-G GGA GAA ATA CAA ATA GGC AGG A…(40n)…TTC GAC AGG AGG CTC ACA ACA GGC -3'
2. 5'- TAA TAC GAC TCA CTA TAG GGA GAA ATA CAA ATA GGC AGG A-3'
3. 5'-GCC TGT TGT GAG CCT CCT GTC GAA-3'

Fig. 1 Sequence 1 represents the ssDNA template with the 5′ conserved region (*red*) and the 3′ conserved region (*blue*) highlighted. Sequence 2 shows the 5′ primer. The 3′ portion (*red*) of the 5′ primer is the exact sequence of the 5′ conserved region of the template; however there are extra bases (*black*) that provide the T7 promoter sequence for later transcription. The third sequence is the 3′ primer, which is the reverse-complement of the 3′ conserved region (when written in the 5′–3′direction)

bind to or mediate the formation of materials [3–6]. In addition to potentially affording more environmentally friendly routes to inorganic materials, the sometimes highly selective recognition capabilities of biomolecules selected in vitro can facilitate their use in fabricating advanced devices such as the viral batteries fabricated by Belcher and coworkers [7].

In this chapter, protocols for preparing and screening large random sequence RNA libraries to isolate specific sequences capable of interacting with inorganic materials are described. Two general examples are provided: (1) isolation of sequences capable of mediating the growth of nanoparticles from solutions containing Co^{2+} and Fe^{2+} precursor ions [8], and (2) isolation of sequences that bind to single-crystal Pt surfaces. When performed successfully, these experiments can provide interesting insights into biomolecule sequence/structure–function relationships in addition to establishing a genomic archive for the synthesis of new materials.

2 Materials

1. A random pool of single-stranded DNA (ssDNA) containing *n* random bases flanked by fixed sequence regions (the ssDNA template). The random region is typically 40–80 nucleotides in length.
2. 3′ and 5′ primers: The primer that anneals to the 3′ end of the template ssDNA must be the reverse complement (when written in the 5′–3′ direction) of the 3′ fixed region of the ssDNA template. The 5′ primer (that which anneals to the 5′ fixed region) must have the same sequence as the 5′ fixed region of the ssDNA template plus additional bases that code for the T7 RNA polymerase promoter. Primer solutions are typically kept as 100 μM aqueous stock solutions at −20 °C. *See* Fig. 1 for a specific example.
3. QIAquick PCR purification kit.
4. Taq DNA polymerase.

5. 10× Taq buffer-$MgCl_2$ (500 mM KCl, 100 mM Tris–HCl, 1 % Triton X-100).
6. 50 mM $MgCl_2$.
7. Dimethyl sulfoxide (DMSO).
8. T7 RNA polymerase.
9. RNase inhibitor.
10. 10 mM dNTPs.
11. 100 mM NTPs.
12. 0.1 M DTT (dithiothreitol).
13. 40 % (19:1) bis-polyacrylamide.
14. *N*,*N*,*N*′,*N*′-Tetramethylethylenediamine (TEMED).
15. 10 % Ammonium persulfate (APS, aqueous).
16. Diethylpyrocarbonate (DEPC).
17. Milli-Q water.
18. Ethanol.
19. 5× Transcription buffer (200 mM Tris–Cl pH 7.5, 6 mM $MgCl_2$, 10 mM spermidine, 100 mM NaCl).
20. 2× denaturing loading buffer (90 % (v/v) formamide, 1 mM EDTA, 0.05 % (w/v) bromophenol blue, 0.005 % (w/v) xylene cyanol).
21. 2× native loading buffer (*see* **item 20**; instead of formamide add 10 % (v/v) of glycerol).
22. 0.22 μm syringe filters and syringes.
23. Ethidium bromide.
24. Polyacrylamide gel (PAGE) equipment.
25. Urea.
26. 1× Wash buffer (1 mM each NaCl, KCl, Na_2PO_4).
27. 10× TBE (electrophoresis buffer).
28. AMV reverse transcriptase (AMV-RT).
29. First Strand Buffer (New England Biolabs).
30. Super Script III Reverse Transcriptase.
31. 1 M Tris–HCl pH 8.0.
32. DNA ladder.
33. RNA ladder.
34. α-^{32}P-ATP.
35. Formamide (molecular biology grade, Sigma).
36. All water used should be high purity (Milli-Q), DEPC treated, and RNase and DNase free.

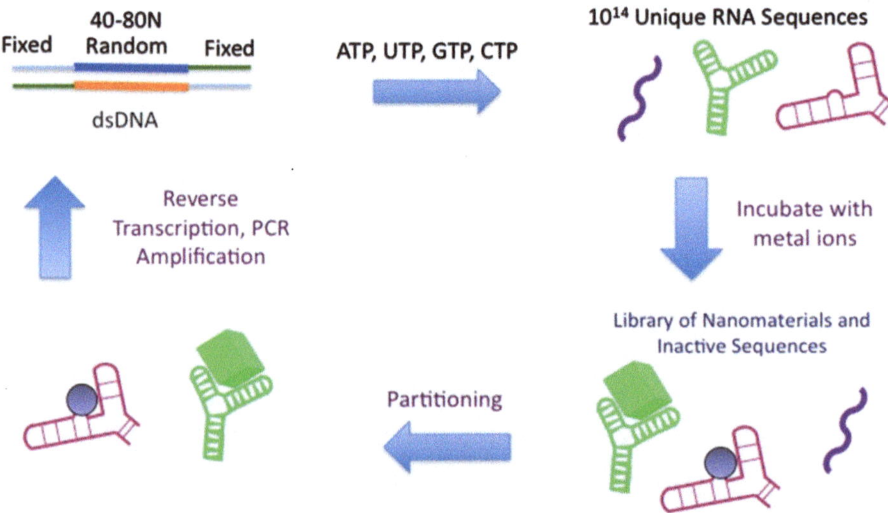

Fig. 2 Basic steps in the in vitro selection of RNA sequences capable of mediating the formation of inorganic nanoparticles

3 Methods

3.1 The Systematic Evolution of Ligands by EXponential Enrichment (SELEX)

RNA sequences with desired function are obtained using multiple cycles of SELEX [9, 10]. SELEX requires the polymerase chain reaction (PCR), transcription, and reverse transcription in combination with a selection pressure and a partitioning method (Fig. 2). The first step in SELEX is 2-cycle PCR that produces double-stranded DNA containing the T7 RNA polymerase promoter sequence. The selection should begin with 1 nmol of ssDNA, which may be obtained from commercial sources (Integrated DNA Technologies is an excellent source). The starting pool of ssDNA will have ca. 10^{14} different dsDNA sequences, which upon enzymatic transcription yields roughly the same diversity of RNA sequences.

1. Start the selection with 2-cycle PCR reaction. Combine in a 250 μL Eppendorf tube the following reagents: 2 μL 10 μM ssDNA random library, 2.5 μL 10 mM dNTPs, 1 μL 100 μM of both 3′ and 5′ primers, 10 μL 10× Taq DNA polymerase buffer, 2.5 U of Taq DNA polymerase, 10 μL DMSO, and 6 μL 50 mM $MgCl_2$. Add DEPC-treated water to obtain a 100 μL final volume. Prepare multiple 100 μL reactions so that 1 nmol of the ssDNA template is subjected to 2-cycle PCR. Place the tubes in a thermal cycler using the following program, repeating **steps** (**b**–**d**) two times:
   (a) 1 min at 95 °C.
   (b) 30 s at 95 °C.
   (c) 30 s at 55 °C*.

(d) 30 s at 72 °C.

(e) Cool to 4 °C.

*The annealing temperature depends on the composition of nucleotides in the primers. Usually this temperature is in the range 50–58 °C. Check this temperature with the vendor providing the primer or calculate the annealing temperature according to the formulas available in molecular biology handbooks. The simplest equation for calculating annealing temperature is the Wallace rule:

$$T_d = 2^{\circ}\mathrm{C}(A+T) + 4^{\circ}\mathrm{C}(G+C)$$

T_d is a filter-based calculation and A, G, C, and T are the number of each base that appears in the sequence. This is often used for shorter (10–20 bases) sequences. Another familiar equation for DNA, which is valid for oligonucleotides longer than 50 bases dissolved in pH 5–9 solutions, is

$$T_\mathrm{m} = 81.5 + 16.6\log M + 41(\mathrm{XG} + \mathrm{XC}) - 500 / L - 0.62F$$

where M is the molar concentration of monovalent cations, XG and XC are the mole fractions of G and C in the oligo, L is the length of the shortest strand in the duplex, and F is the molar concentration of formamide.

2. DNA purification. Purify the double-stranded DNA (dsDNA) random library after the 2-cycle PCR reaction using the Qiagen PCR purification kit, following the Qiagen protocol exactly. Verify that the dsDNA pool is of the correct length using native PAGE (typically between 6 and 8 % depending upon the DNA length) by comparison with appropriate DNA ladder. If the dsDNA is not of the appropriate length, or multiple products are observed, the dsDNA must be purified by native PAGE.

3. T7 RNA transcription. Convert the dsDNA random pool into radiolabeled RNA using enzymatic transcription in the presence of α-^{32}P-ATP. Prepare ten reactions containing the following: 0.1 nmol of random dsDNA, 1 mM of each NTP, 100 U T7 RNA polymerase, 20 μL 5× transcription buffer, 100 U RNase inhibitor, 15 μCi α-^{32}P-ATP (for a 100 μL reaction)**, and 10 μL 100 mM DTT. Add water to 100 μL. Incubate the reactions at 37 °C for 6 h.

**Calculate the specific activity using the following formula and the date on the bottle:

$$\mathrm{SA}(\mathrm{Ci} / \mathrm{mol}) = 3{,}000\mathrm{e}^{-(K_\mathrm{decay} \times t)}$$

If ATP is dated the first and it is the fifth, t=4, and if it is dated the fifth and it is the first, then t=–4. Notice the following:

3,000 is the maximum activity of μCi α-^{32}P, and k_{decay} for α-^{32}P is 0.0485/day. Use the SA to calculate the volume of μCi α-^{32}P-ATP:

$$\text{vol}(\mu\text{L}) = (15\mu\text{Ci}) \times (1 / \text{SA}) \times (1 / [\alpha\text{-}^{32}\text{P-ATP}])$$

The RNA should be checked for purity via denaturing PAGE (8 %). Purify the random RNA library using 8 % denaturing PAGE. Quantitation can be performed on radiolabeled RNA using a liquid scintillation counter or by measuring the OD_{260} of the RNA solution.

The readout from your liquid scintillation counter will likely be in counts per minute (CPM). CPM will need to be converted into moles. Calculate a new SA if the previous SA was from a different day. Calculate the hotness factor (HF):

$$\text{HF} = \left(\frac{\text{moles ATP added}}{\text{moles } \alpha\text{-}^{32}\text{P-ATP added}}\right)(\text{fraction of ATP in sequence})$$

The following equation can be used to convert CPM to moles:

$$\left[\frac{x\text{counts}}{1\,\text{min}} \times \frac{1\,\text{min}}{60\,\text{s}} \times \frac{1\,\text{s}}{3.7\text{E}10\text{ decays}} \times \frac{1}{\text{SA}} \times \frac{\text{total vol RNA}}{(\text{vol RNA added} \times \text{vol cocktail})} \times (\text{HF})\right] = \text{moles RNA}$$

The purified RNA is now ready to be used in the desired selection.

3.2 Partitioning

3.2.1 Partitioning for a Desired Property

The partitioning step in SELEX depends upon the desired outcome. For instance, if high-affinity RNA binders to a certain inorganic crystal facet are desired, simple rinse protocols may be used to partition the weaker-binding sequences from the stronger binders (*see* below). Particle size may also serve as a selection criterion, and filtration membranes and/or gels may be employed to partition sequences that mediate the formation of and bind to inorganic nanoparticles. Here we illustrate a selection protocol designed to isolate RNA sequences capable of mediating the formation of cobalt iron oxide nanoparticles [8]. The partitioning step in this example was performed with application of a magnetic field. Nanoparticles and their RNA cognates that responded to the magnetic field were partitioned from inactive sequences.

1. Incubation of RNA with the metal precursors. Combine the following reagents in a 1 mL Eppendorf tube: 900 nM RNA, 100 mM NaCl, 100 mM KCl, 10 mM HEPES (pH 7.2), 150 μM degassed $FeCl_2$, and 75 μM degassed $CoCl_2$. Add H_2O to 400 μL of final volume. Allow the reaction to incubate at room temperature for 5 h.
2. Immediately prior to partitioning, remove the entire reaction solution from the Eppendorf tube and place it in a new tube to

eliminate any RNA sequences that were bound nonspecifically to the tube.

3. Partition. Place the tube in a magnetic field (permanent magnet) for 12 h at 4 °C.
4. Remove the solution from the tube, being careful not to touch the sides of the tube with the pipette tip. Add 200 μL wash buffer and remove with a pipette, and repeat washing four times. Remove the tube from the magnetic field and resuspend the RNA sequences remaining in the tube in 100 μL H_2O.
5. Quantify the amount of resuspended RNA via liquid scintillation and calculate the percent of RNA resuspended after the partitioning step. The percent RNA retained typically increases with each cycle unless a selection pressure has been changed.
6. Perform the reverse-transcription on the recovered RNA. Combine the resuspended RNA from **step 3**, 5× First Strand Buffer; 0.5 mM each dATP, dCTP, dGTP, and dTTP; 2 μM 3′ primer; and 0.2 U/μL AMV reverse transcriptase. Incubate the reaction at 42 °C for 45 min followed by 15 min at 72 °C (to inactivate the enzyme).
7. Skip to Subheading 3.3.

3.2.2 Selection of RNA Sequences That Bind to a Pt Surface

The selection of RNA sequences bound to a metal surface is illustrated here for two single-crystal Pt surfaces (Fig. 3). The Pt crystal plates with exposed (100) and (117) facets were purchased from MaTeck Gmbh. The plates were 1 mm thick with a diameter of 25 mm. A specialized reactor fabricated from Kel-f was designed to control the amount of surface area of Pt that was available to be bound by the RNA library. The design was based upon work by J. E. Hutchison and coworkers. The reactor, shown in Fig. 4, has a 3 mm diameter hole at the bottom allowing for the RNA sequences to come in contact with the Pt surface. The bottom surface of the reactor was highly polished so that a tight seal could be made between the reactor and the Pt surface. The reactor was sealed to the Pt surface and a solution containing the RNA library was placed inside. The key to the selection process is the application of a selection pressure, in this case applying pressure, so that the highest-affinity Pt-binding RNA sequences were obtained. This was done by decreasing the RNA concentration and surface area of the Pt wafer as the selection progressed. Decreasing the available area of Pt was accomplished by decreasing the diameter of the bottom hole of the reactor:

1. Place the plastic reactor on the platinum plate; screw the cover to sill the reactor to the plate, place the small stir bar inside the reactor, and place the reaction vessel on the stirring/heating plate. Wash three times the reaction vessel with the 200 μL of buffer containing 20 mM HEPES pH 7.1, 5 mM $MgCl_2$, and 30 mM NaCl with stirring 5 min for each wash at room temperature.

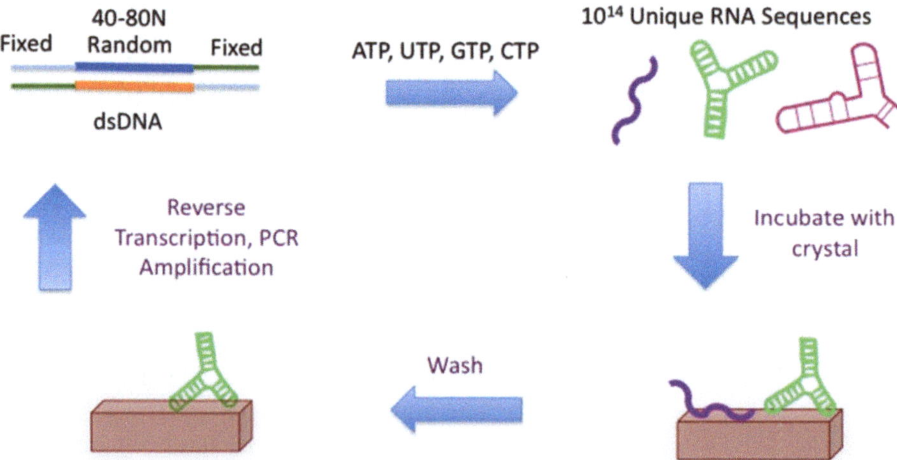

Fig. 3 Basic steps in the in vitro selection of RNA sequences capable of binding to solid substrates

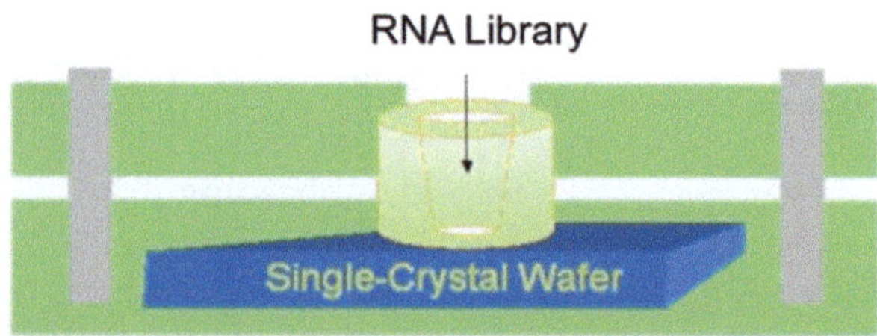

Fig. 4 Schematic of a cell designed to select RNA sequences capable of binding to single-crystal wafers

2. Heat the reaction mixture containing 1 nmol of RNA with 40-nucleotide-long random region in 20 mM HEPES pH 7.1, 5 mM $MgCl_2$, and 30 mM NaCl in the total volume of 200 μL for 5 min at 95 °C. Then slowly cool down the mixture on ice.
3. Place the cooled RNA solution into the reaction vessel and incubate the RNA on the platinum surface for 1 h at room temperature with mixing.
4. Discard unbound RNA sequences by removing the reaction solution from the reaction vessel.
5. Wash the reaction vessel three times with the reaction buffer from **step 1**.
6. Elute the RNA sequences bound to the platinum surface with 200 μL of formamide (deionized, molecular biology tested, SIGMA). Elute bound RNA sequences by stirring for 1 h at 75 °C. Pipette the eluate into a microcentrifuge tube. Repeat **step 6**.
7. Precipitate the eluted RNA sequences by adding 20 μL of 3 M sodium acetate (pH 5) and 600 μL ethyl alcohol to each tube. Allow the RNA to precipitate at −80 °C for at least 2 h, preferably overnight.

8. Spin the ethanol solution at 16,363 × *g* for 15 min at 4 °C. Decant and discard the supernatant, and remove the residual of supernatant.
9. Wash the pellet with 300 μL of 70 % ethanol and vortex. Spin at 13,200 RPM for 15 min. Discard the supernatant and remove the residual alcohol with the pipette. Leave the tube open for 15 min to dry the pellet in air. Dissolve the pellet in 20 μL of DEPC-treated water.
10. Reverse transcribe the eluted RNA sequences. Prepare a solution containing (Solution A): 20 μL of RNA from **step 9**, 400 pmol of 3′ primer (4 μL of 100 μM solution), and 38 μL of DEPC-treated water. Prepare a second solution (Solution B) containing the following: 20 μL 5× Super Script III reaction buffer, 5 μL 100 mM DTT, 10 μL 10 mM dNTPs, and 3 μL Super Script III reverse transcriptase. Transfer Solution A into a 200 μL PCR tube and place it in the thermocycler and apply the following program:
    (a) 5 min at 85 °C.
    (b) 5 min at 42 °C.
    (c) 30 min at 55 °C.
    (d) 20 min at 37 °C.
    (e) 15 min at 75 °C.
    (f) Hold at 4 °C.
11. When the temperature reaches 42 °C add Solution B. When the temperature reaches 37 °C add 1 μL of RNaseH and continue the program to the end.

3.3 DNA Amplification (PCR)

Perform PCR pilots to determine the proper number of cycles of PCR that are required to maximize dsDNA yield while maintaining the proper length. Combine in a 250 μL Eppendorf tube the following reagents: 5–10 μL cDNA from Subheading 3.2.2, **step 11**, 0.1 U/μL Taq DNA polymerase, 10 μL 10× Taq DNA polymerase buffer, and 0.12 mM each dATP, dCTP, dGTP, and dTTP. Add water to final volume of 100 μL. Place the tube in a thermal cycler and apply following program:

(a) 2 min at 95 °C.
(b) 30 s at 95 °C.
(c) 30 s at 60 °C.
(d) 30 s at 72 °C.
(e) Cool to 4 °C.

Repeat **steps** (**b–d**) for 30 cycles, taking an aliquot of the PCR reaction every five cycles for gel analysis. Collect ~2 μL aliquots of the PCR reaction at 5, 10, 15, 20, 25, and 30 cycles and run native-PAGE to determine the appropriate number of cycles. The appropriate number of cycles will maximize the dsDNA of the correct length while minimizing higher molecular products. Repeat the PCR with the determined number of cycles with several 100 μL reactions to obtain dsDNA to carry forward to the next cycle of selection. Purify the dsDNA using the QIAquick PCR purification kit. Dissolve the amplified DNA in 300 μL of water.

3.4 T7 RNA Transcription

Repeat transcription as described in Subheading 3.1, **step 3**. We recommend making 5–100 μL reactions with 5–10 μL of amplified DNA from Subheading 3.3. After 6 h incubation remove excess α-^{32}P-ATP, NTPs, and reaction buffer using a molecular weight cutoff filter (Microcon 10).

Prewash the filter with 200 μL of 1× wash buffer (Subheading 2, **step 26**), for 5 min at 13,200 RPM. Add the transcription reactions, and concentrate at 13,200 RPM for 5 min. Wash the reaction on the Microcon filter 4× with 200 μL 1× wash buffer, spinning for 5 min at 13,200 RPM. To recover the RNA product, resuspend in 50–100 μL H_2O, invert into a clean tube, and spin at 4,000 RPM for 5 min. (Instructions are provided with the Microcon filters.) RNA purity should be confirmed via 6 % denaturing PAGE. (Gel purification may be required).

3.5 Analysis and Unique Identification of Selected RNA Sequences

The selected RNA sequences are analyzed and separated via cloning and sequencing. The evolved pool of RNA obtained from the final cycle of SELEX is reverse transcribed and then amplified by PCR. The "final cycle" is somewhat arbitrary; however, a selection typically requires between 8 and 12 cycles with large increases in the percent RNA obtained indicating convergence on specific RNA sequences capable of performing the desired task.

1. Cloning can be performed using a PCR-Script Amp Cloning Kit from Stratagene (or an analogous kit), following the provided protocol exactly.
2. The plasmid is digested and individual sequences from the evolved pool are ligated into individual plasmids.
3. The plasmids are then transformed into competent cells.
4. The competent cells are plated and grown on ampicillin-resistant LB-Agar with Xgal/IPTG for blue/white screening. White colonies contain the inserted sequence from the evolved pool.
5. Individual colonies are picked and grown in LB containing ampicillin. Growing the individual colonies amplifies the individual sequence from the evolved RNA pool. The cells can then be pelleted at 5000×*g*, decanted, and stored at −20C until they can be sent for sequencing.
6. Sequencing of the individual sequences from the evolved pool.
   (a) Sequencing by Bio-Basic or similar contract lab.
   (b) The obtained sequences may be analyzed by free software available online.

Table 1
Kinetic data for a single RNA isolate, the initial random RNA sequence library, and the evolved RNA pool

	Random library	Cycle 7 pool	14	26	3
k_{ads} (s^{-1})	0.0046	0.008	0.030	0.010	0.006
stdev	0.0009	0.007	0.006	0.007	0.004

(c) Sequence similarity and consensus regions may be identified and analyzed providing insight into sequence–function relationships of the selected RNA sequences.

3.6 Characterization

3.6.1 Characterization of the Binding Kinetics and Thermodynamics of Selected RNA Sequences

1. The adsorption rate constant (k_{ads}) can be determined for any single RNA sequence, the random RNA pool, and the evolved pool of RNA. For the selection of RNA sequences that bind to Pt, a constant amount of ^{32}P-labeled single sequence RNA (2.5 pmol in 2 μL) was dissolved in the reaction buffer (*see* Subheading 3.2.2, **step 2**) and incubated on the single-crystal Pt plate for varying amounts of time (5, 15, 30, 60, 150, 300 s). The Pt plate was then rinsed with mQ water, dried in air, and exposed to a phosphor screen (Packard). The amount of RNA bound to the Pt plate was quantified (Packard Cyclone) and analyzed by OptiQuant software. The k_{ads} values were calculated using the following formula for the pseudo first-order adsorption at a surface with no site–site interactions:

   The values of k_{ads} are shown in Table 1.

   Only sequence 14 had a statistically larger adsorption rate constant compared to the random RNA library based upon Student's *t*-test at the 95 % confidence level. This suggests that the selection pressure was likely not kinetic but thermodynamic in origin. This was examined further by measuring the binding affinities of the selected sequences.

2. To characterize the binding affinities of the selected RNA sequences, the adsorption affinity constant, k_{ads}, was determined. Varying amounts of ^{32}P-labeled RNA (3.8 nM, 7.5 nM, 14 nM, 80 nM, 160 nM, 500 nM, 750 nM, and 1.0 μM) were dissolved in the reaction buffer (*see* Subheading 3.2.2, **step 1**) and incubated for 1 h on the single-crystal Pt plate. The plate was then rinsed with mQ water, dried in air, and exposed to the phosphor screen. The amount of RNA bound to the Pt plate was quantified and analyzed by OptiQuant software using a Packard Cyclone. The k_{ads} values were calculated using the Langmuir isotherm equation:

$$\frac{c}{\Gamma} = \frac{c}{\Gamma_{max}} + \frac{1}{K_{eq}\Gamma_{max}}, \qquad \Gamma = \frac{[\text{counts}]_c}{[\text{counts}]_{csat}}$$

   The values of k_{ads} for individual RNA sequences are shown in Table 2.

Table 2
Pt binding constants for random RNA library and selected RNA sequences

	Random library	14	26	3	8
K_{ads} (×10^7)	0.3	1.3	3.7	0.8	0.9
Stdev (×10^7)	0.07	0.3	0.8	0.3	0.2

The values of k_{ads} for sequences 14, 26, and 8 were statistically different from the random RNA library based upon Student's *t*-test at the 95 % confidence level. The higher affinity of the selected RNA sequences for the Pt surfaces suggests that they responded to a thermodynamic selection pressure in a way that afforded them an advantage over other members of the random RNA library. Additional cycles of selection or a more stringent selection pressure would be expected to yield sequences with even higher affinities.

3.6.2 Nanoparticle Characterization

1. Transmission electron microscopy (TEM). TEM can be used to image particles as small as 3–5 nm. In addition to the ability to image nanoscale materials, relatively small (μL) quantities of sample are required. Diffraction patterns and elemental analysis are also readily performed with most transmission electron microscopes. General sample preparation protocol using carbon-coated copper grids (Ted Pella):
   (a) Place a glass microscope slide into the bottom half of a glass petri dish.
   (b) Using fine tip tweezers, place a grid on the edge of the glass carbon side up.
   (c) Glow discharge the grid to apply a charge to the carbon surface making it hydrophilic.
   (d) Remove the grid from the slide and allow it to remain clamped in the tweezers.
   (e) Apply a 2–20 μL drop of the sample onto the grid allowing it to adsorb for 30 s to 2 min.
   (f) Wick away the solution using a piece of filter paper by gently touching the paper to the side of the grid.
   (g) If desired, a 20 μL drop of water can be applied to the grid and wicked away to remove salts present in buffers used during synthesis.
   (h) Allow the grid to dry in air, covering with a petri dish lid to avoid contamination, for at least 15 min.
   (i) Sample image obtained using this method is shown in Fig. 5.

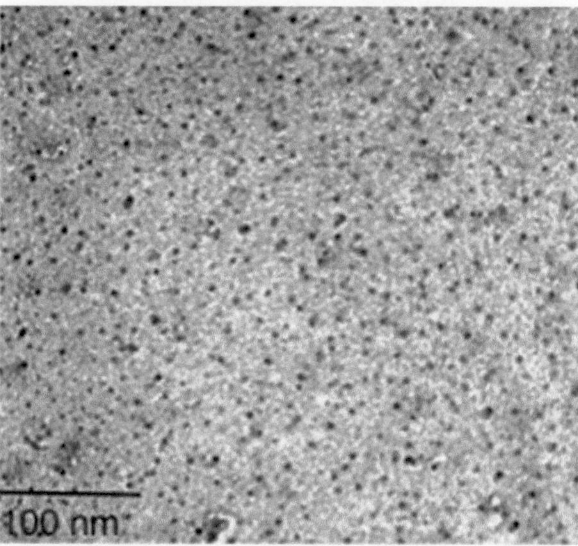

Fig. 5 Sample transmission electron microscope image of Co Fe oxide nanoparticles synthesized with a selected RNA sequence

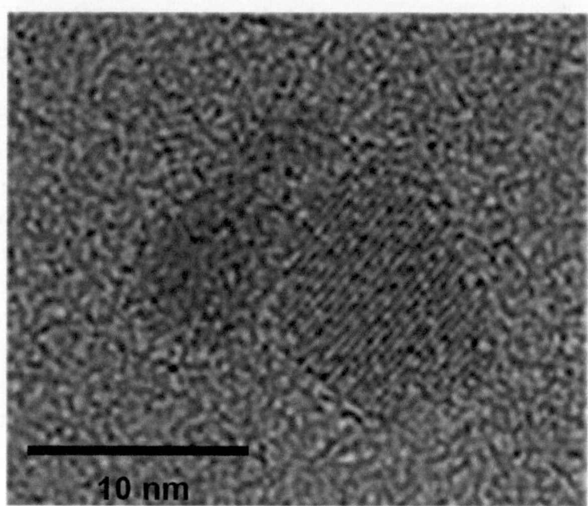

Fig. 6 HR-TEM image of a cluster of 3–10 nm diameter iron oxide particles

2. High-resolution transmission electron microscopy (HR-TEM): High-resolution microscopes provide the capability of imaging features on the order of angstroms. Lattice spacings of crystalline or polycrystalline nanomaterials can be obtained from the images, providing another method for materials characterization. Figure 6 shows an image of a metal oxide particle taken using HR-TEM.
3. Dynamic light scattering (DLS): DLS uses changes in the intensity of input light compared to output light to determine the average size and dispersity of a nanoparticle solution.

4. Elemental analysis tools useful in nanoparticle characterization:
   (a) Electron energy loss spectroscopy (EELS).
   (b) Energy-dispersive X-ray spectroscopy (EDS).
   (c) X-ray photoelectron spectroscopy (XPS).
   (d) Selected area electron diffraction (SAED).

4 Notes on Optimization and Troubleshooting

4.1 How Many Cycles of SELEX Do We Need to Perform?

The selection of active RNA sequences requires repeating selection cycles several times to obtain the best sequences. A typical RNA selection requires 8–12 cycles of SELEX. To select RNA sequences that bound Pt 100 or 117, seven cycles of SELEX were performed. We note that increasing the number of cycles and selection pressure would likely result in higher affinity binders than those shown in Table 2.

4.2 How Do I Increase the Selection Pressure?

Applying a selection pressure can be accomplished by decreasing the concentration of RNA and/or the concentration of the target material prior to partitioning. The incubation time can also be decreased to apply a selection pressure.

4.3 No PCR Product or the Wrong Length of PCR Product?

After each round of PCR, the purity and length of DNA should be verified via native-PAGE. If low concentrations of RNA sequences were bound to the target and/or not eluted efficiently from the target, reverse transcription may be inefficient. Low concentrations of cDNA from RT could cause unproductive PCR. While it may be counterintuitive, increasing the number of PCR cycles does not solve the problem. The last round of SELEX needs to be repeated.

4.4 How Can I Elute Bound RNA from a Surface?

Eluting the bound RNA from a surface is a crucial step in isolating high-affinity binders. The elution should release the RNA from the target but must not result in RNA degradation. Denaturing reagents such as formamide, 8 M urea, and DMSO are useful in combination with elevated temperature. However, heating RNA to more than 75 °C will result in RNA degradation.

References

1. Aizenberg J, Tkachenko A, Weiner S, Addadi L, Hendler G (2001) Calcitic microlenses as part of the photoreceptor system in brittlestars. Nature 412:819–822
2. Blakemore R (1975) Magnetotactic bacteria. Science 190:377–379
3. Richards CI et al (2008) Oligonucleotide-stabilized Ag nanocluster fluorophores. J Am Chem Soc 130:5038
4. Sewell SL, Wright DW (2006) Biomimetic synthesis of titanium dioxide utilizing the R5 peptide derived from Cylindrotheca fusiformis. Chem Mater 18:3108–3113
5. Whaley SR, English DS, Hu EL, Barbara PF, Belcher AM (2000) Selection of peptides with semiconductor binding specificity for directed nanocrystal assembly. Nature 405: 665–669

6. Eaton BE, Feldheim DL (2007) In vitro selection of biomolecules capable of mediating the formation of materials. ACS Nano 1: 154–159
7. Nam KT et al (2008) Stamped microbattery electrodes based on self-assembled M13 viruses. Proc Natl Acad Sci USA 105:17227–17231
8. Carter CJ et al (2009) In vitro selection of RNA sequences capable of mediating the formation of iron oxide nanoparticles. J Mater Chem 19:8320–8326
9. Tuerk C, Gold L (1990) Systematic evolution of ligands by exponential enrichment-RNA ligands to bacteriophage-T4 DNA polymerase. Science 249:505–510
10. Ellington AD, Szostak JW (1990) In vitro selection of RNA molecules that bind specific ligands. Nature 346:818–822

Chapter 11

Single-Walled Carbon Nanotube-Mediated Small Interfering RNA Delivery for Gastrin-Releasing Peptide Receptor Silencing in Human Neuroblastoma

Jingbo Qiao, Tu Hong, Honglian Guo, Ya-Qiong Xu, and Dai H. Chung

Abstract

Small interfering RNA (siRNA) has the potential to influence gene expression with a high degree of target gene specificity. However, the clinical application of siRNA therapeutics has proven to be less promising as evidenced by its poor intracellular uptake, instability in vivo, and nonspecific immune stimulations. Recently, we have demonstrated that single-walled carbon nanotube (SWNT)-mediated siRNA delivery can enhance the efficiency of siRNA-mediated gastrin-releasing peptide receptor (GRP-R) gene silencing by stabilizing siRNA while selectively targeting tumor tissues. Based on our recent findings, we introduce a novel technique to silence specific gene(s) in human neuroblastoma through SWNT-mediated siRNA delivery in vitro and in vivo.

Key words SWNT, siRNA, GRP-R, Delivery, Image, Neuroblastoma

1 Introduction

Neuroblastoma is the most common extracranial solid tumor in infants and children. Despite recent advances in multimodality therapy, survival rates for all stages of tumors remain a dismal 50 %. Novel therapeutic options are, therefore, urgently needed to improve patient outcomes [1]. Gastrin-releasing peptide (GRP), the mammalian equivalent of bombesin, is a neuroendocrine peptide that has been shown to have a growth stimulatory effect on various cancer cell types [2–4]. We have previously demonstrated that GRP and its cell surface receptor, GRP-R, are abundantly expressed in human neuroblastomas [2, 5] and that GRP-R silencing significantly suppresses tumorigenesis and metastatic potential in mouse tumor models of neuroblastoma [6]. To employ siRNA, we relied on the silencing of GRP-R by techniques previously described in the literature [6].

However, current applications of siRNA therapeutics are limited by the lack of effective delivery systems. Therefore, an efficient and

Sandra J. Rosenthal and David W. Wright (eds.), *NanoBiotechnology Protocols*, Methods in Molecular Biology, vol. 1026, DOI 10.1007/978-1-62703-468-5_11,

safe targeted silencing delivery system is highly desired. Recently, single-walled carbon nanotube (SWNT)-mediated delivery systems have garnered significant interest as a novel tool for drug delivery. This is, in large part, due to the ability of SWNTs to cross cell membranes and transport a wide range of biologically active molecules including drugs, proteins, DNA, and RNA into cells [7–9]. The cytotoxicity of SWNTs is mainly dependent on their surface functionalization, where minimal toxic effects were seen in well-functionalized, serum-stable SWNTs. For example, phospholipid-polyethylene glycol (PL-PEG)-coated SWNTs have been shown to be nontoxic and are widely used in vitro and in vivo [10].

Moreover, SWNTs have emerged as one of the most promising candidates in the field of diagnostic imaging due to their unique optical properties [11–15]. The emission wavelength of SWNTs is in the near-infrared (NIR) range, where most biological tissues and cells are transparent and have low inherent autofluorescence. The fluorescence of SWNTs is extremely stable and can last for months, while fluorescent proteins, which are currently used for imaging living systems, begin to fade within hours. Thus, SWNTs may have significant implications in how we approach fluorescent visualization.

With the capability of penetrating cells and emitting non-bleaching NIR fluorescence, functionalized SWNTs offer the potential to be used as carriers for the delivery of siRNA and diagnostic imaging agents in vivo. To address the aforementioned application, we have developed a straightforward protocol for silencing GRP-R in human neuroblastoma through SWNT-mediated siRNA (SWNT–siRNA) delivery in vitro and in vivo. Western blotting and immunohistochemistry techniques have been used to confirm GPR-R silencing.

2 Materials

2.1 Single-Walled Nanotubes (SWNTs)

1. High-Pressure CO conversion (HiPco) SWNTs.

2.2 Chemicals

1. PL-PEG (1,2-distearoyl-*sn*-glycero-3-phosphoethanolamine-*N*-[methoxy(polyethylene glycol)-5000]) (Avanti Polar Lipids Inc).
2. DTT (D,L-dithiothreitol) (Sigma-Aldrich).

2.3 Biological Reagents

1. Antibodies:
   (a) GRP-R primary antibody (Abcam).
   (b) ß-actin antibody (Sigma-Aldrich).
   (c) Secondary antibodies against mouse and rabbit IgG (Santa Cruz Biotechnology, Inc.).
   (d) Rabbit polyclonal anti-human phospho-histone H3 (Ser10) antibody (Millipore).

(e) Alexa Fluor 568 Goat Anti-Rabbit IgG (Life Technologies).

2. siRNAs were custom synthesized by Thermo Scientific (Dharmacon):
   (a) Custom siRNA Design Tool, Thermo Scientific siDESIGN Center, http://www.dharmacon.com/PopUpTemplate.aspx?id=2078
   (b) The siRNA targeting GRP-R (NM_005314) sequence:
      - Sense: 5′-thiol-UAACGUGUGCUCCAGUGGAdTdT-3′
      - Antisense: 3′-dTdTAUUGCACACGAGGUCACCU-5′
   (c) Luciferase siRNA (control):
      - Sense: 5′-thiol-CUUACGCUGAGUACUUCGAdT dT-3′
      - Antisense: 3′-dTdTGAAUGCGACUCAUGAAGCU-5′
   (d) siRNA solution can be prepared by dissolving siRNA in RNase-free water to reach a siRNA concentration of 100 μM. Store aliquots of the siRNA solution at −80 °C, which can be stable for up to 6 months. However, avoid more than three freeze–thaw cycles.
3. Cell-Counting Kit-8 (Dojindo Molecular Technologies, Inc.).
4. Cell lysis buffer (10×) (Cell Signaling Technology).
5. Protein inhibitors cocktail (Roche Applied Science).
6. NuPAGE Novex 4–12 % Bis–Tris Gel (Invitrogen).
7. NuPAGE MOPS Running buffer (20×) (Invitrogen).
8. TG (10×) buffer (Bio-Rad).
9. Western Lightning Plus-ECL, enhanced Chemiluminescence substrate (PerkinElmer, Inc.).
10. X-Ray Film, Blue Basic Autorad Film (ISC BioExpress).
11. Immun-Blot PVDF Membrane (Bio-Rad).
12. Agarose gels (Cambrex Bio Science Rockland, Inc., Rockland, ME).
13. DAPI (4′,6-Diamidino-2-Phenylindole, Dihydrochloride, Sigma-Aldrich).
14. DAKO EnVision + System-HRP (Dako North America, Inc., Carpinteria, CA).

2.4 Cell Culture Reagents

1. The human neuroblastoma cell lines SK-N-SH and BE(2)-C (American Type Culture Collection).
2. RPMI 1640 medium and 0.25 % trypsin (Mediatech, Inc.).
3. Fetal bovine serum (FBS) (Sigma-Aldrich).

4. Phosphate-buffered saline (PBS), 10×, pH 7.4 (Invitrogen).
5. Penicillin–streptomycin, liquid (10,000 U penicillin; 10,000 μg streptomycin) (Invitrogen, GIBCO).

2.5 Instruments

1. Bath sonicator, Branson 2510.
2. Cole Palmer CPX-600 cup-horn sonicator.
3. Polyscience X-520 Homogenizer, 750 W.
4. Sorvall Legend X1R Centrifuge (Thermo Scientific); Rotor, FiberLite, F15-8X50C (Piramoon Technologies Inc.).
5. Sorvall WX Ultra Centrifuge with Surespin 630 swing bucket rotor (Thermo Scientific).
6. Dialysis tubing 15 ml, 3,500 MWCO (Fisher Scientific).
7. Amicon centrifugal filter device 4 ml, 100,000 MWCO (Millipore).
8. Microcon Ultracel YM-100 filter device 0.5 ml, 100,000 MWCO (Millipore).
9. Illustra NAP-5 columns, Sephadex G-25 DNA Grade (GE Healthcare).
10. Cary-5000 UV–VIS–NIR Spectrometer (Varian).
11. Cell culture plates and flasks (BD Falcon).
12. Cell culture incubator, Thermo Series II, water Jacketed CO_2 Incubator, HEPA class 100 (Thermo Electron Corporation).
13. FlexStation 3 (Molecular Devices Corp.).
14. Immuno blotting system (Invitrogen).
15. Mini Trans-Blot Electrophoretic Transfer Cell (Bio-Rad).
16. Nikon Eclipse E600.
17. Leica DMI6000 B.
18. Inverted Microscope IX 71 (Olympus).
19. IR enhanced water immersion objective (Olympus).
20. Liquid N_2 Cooling 2D OMA-V 320 Detector (Princeton Instruments).
21. ProEM 512B CCD (Princeton Instruments).

3 Methods

3.1 Preparation of Functionalized SWNT

1. Weigh 1 mg of HiPco SWNTs and 5 mg of PL-PEG into a 20 ml glass scintillation vial. Add 5 ml of water. Dissolve PL-PEG completely by shaking.
2. Sonicate the vial in an ice water bath sonicator for 120 min and change the ice water every 30 min to avoid overheating.

3. Centrifuge the SWNT suspension at 133,000 × *g* and at room temperature for 4 h, and then decant the upper 75–80 % of the supernatant.
4. Measure the concentration of the prepared PL–PEG functionalized SWNT (PEG–SWNT) solution using a UV–VIS–NIR spectrometer with a weight extinction coefficient of 0.0465 l/mg/cm at 808 nm. The final SWNT concentration normally ranges from 20 to 30 mg/l. Store the PEG–SWNT solution at 4 °C.
5. Add 4 ml of the PEG–SWNT solution from the prepared stock in **step 3** into a 4 ml Amicon centrifugal filter device with a molecular weight cutoff (MWCO) of 100 kDa. Centrifuge the device at 4,000 × *g* and at 4 °C for 10 min. The leftover volume in the filter should be <0.5 ml. Fill the filter device with water to 4 ml. Wash five to six times by repeating the centrifuge and water-adding steps in order to completely remove excess PL-PEG in the PEG–SWNT solution. After the final wash, resuspend PEG–SWNTs in PBS and adjust the concentration of the SWNT solution to ~40 mg/l by adding the required amount of water.

3.2 siRNA Conjugation to SWNTs Through Natural π–π Stacking

1. Dissolve siRNA in RNase-free water to reach a siRNA concentration of 100 μM. Dilute siRNA in PBS to a concentration of 3.4 μM.
2. To conjugate siRNA to functionalized SWNT, mix PEG–SWNT solution from Subheading 3.1 with the siRNA solution in equal volumes and incubating at 4 °C for 24 h. The final SWNT and siRNA concentrations were approximately 20 mg/l and 1.7 μM, respectively. The SWNT–siRNA solution was ready for cell transfection.

3.3 Cell Culture and Transfection with SWNT–siRNA

1. Culture SK-N-SH and BE(2)-C cells in RPMI 1640 medium supplemented with L-glutamine and 10 % FBS. Maintain at 37 °C in a humidified atmosphere of 95 % air and 5 % CO_2.
2. For the siRNA transfection, plate cells in a 6-well plate at a density of 5×10^5 cells/well in 2 ml medium. Incubate cells overnight under the condition described in **step 1**.
3. To remove aggregates, centrifuge the SWNT–siRNA solution prepared in Subheading 3.2 at 10,000 × *g* and 4 °C for 10 min. Collect the SWNT–siRNA supernatant containing either SWNT–siGRP-R or SWNT–siLuc conjugates, and add 500 μl to each well containing cultured cells and 2 ml of medium. The final SWNT and siRNA concentrations should be approximately ~4 mg/l and ~300 nM, respectively. Incubate the cells at 37 °C, 5 % CO_2 overnight.
4. To examine protein expression levels, replate the transfected cells from **step 3** into 6-well plates. Simply remove culture medium and treat the cells with 300 μl of 0.25 % trypsin for

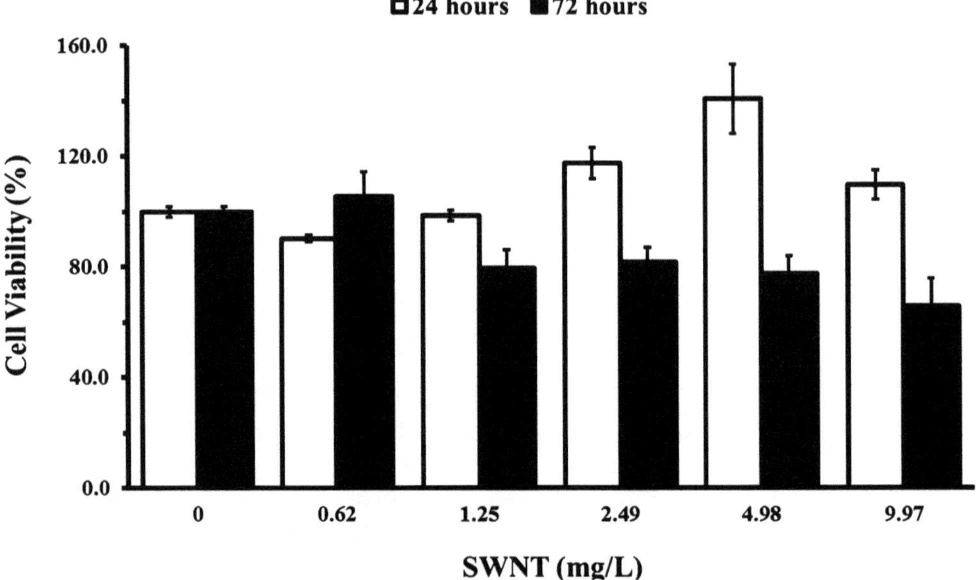

Fig. 1 PEG–SWNT toxicity assay. Cells treated with PEG–SWNT did not exhibit a significant inhibition on cell proliferation when compared to a control (no SWNT treatment)

2–5 min. Then add 1 ml of RPMI medium with 10 % FBS and spin down the cells at 200 × *g* for 5 min at 4 °C. Resuspend and plate the cells into 35 mm dishes for protein extraction at various time points.

3.4 Cell Cytotoxicity Assay

1. Plate cells in 96-well plates at a density of 5–10 × 10^3 cells/well in RPMI 1640 culture medium with 10 % FBS and incubate overnight under the condition described in Subheading 3.3, **step 1**.
2. Add PEG–SWNT solution into the well at various concentrations, and culture these cells for up to 3 days after incubation with SWNTs.
3. Assess the cell numbers using Cell-Counting Kit-8 daily. The values, corresponding to the number of viable cells, can be read at OD450 with FlexStation 3. Set up each assay point in triplicate, and repeat the experiment three times for each cell line (Fig. 1).

3.5 Cell Preparation for NIR Fluorescence Imaging

1. Trypsinize and wash cells after SWNT–siRNA treatments using culture medium via centrifugation at 200 × *g* and at room temperature for 5 min.
2. Resuspend cells at the concentration of 2 × 10^6 cells/ml in RPMI 1640 medium without phenol red.
3. Mix the cell solution with equal volume of 0.8 % agarose gels (warmed at 40 °C) to form a final semisolid cell solution with concentration of 1 × 10^6 cells/ml in 0.4 % agarose gels.
4. Spread the cell gel solution on the glass slide and cover it with coverslip.

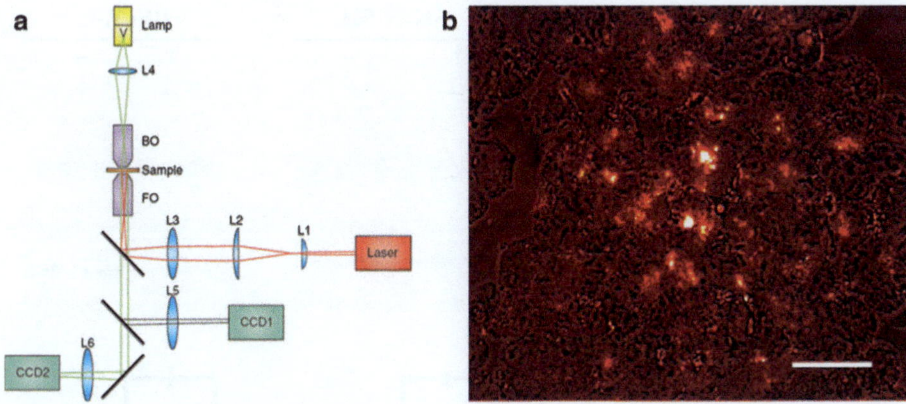

Fig. 2 (**a**) A schematic diagram of the fluorescence imaging setup; (**b**) a NIR fluorescence image of SWNT–siRNA-treated cells overlapped with the related optical image. The scale bar is 40 μm

3.6 NIR Fluorescence Imaging of SWNT–siRNA-Treated Cells

A 70 mW/785 nm laser beam (CrystaLaser) is expanded by lenses 1 and 2 and then focused by lens 3 to the back focal plane of an IR-enhanced 60× water immersion objective (Olympus; Fig. 2a). The laser spot in the sample plane is approximately 100 μm in diameter. NIR fluorescence imaging is carried out in epifluorescence mode, and fluorescence images (Fig. 2b) are collected by a liquid nitrogen cooled two-dimensional InGaAs array (2D-OMA V; Princeton Instruments). The excitation light is filtered out using one 1,150 nm long pass filter (Thorlabs). The visible images are collected using an EMCCD (ProEM 512B CCD; Princeton Instruments).

3.7 Protein Western Blotting for GRP-R

1. Prepare whole cell lysate by mixing cells with lysis buffer (1×) containing 1 mM PMSF and protein inhibitor cocktail.
2. Measure the protein concentrations using Bio-Rad Protein Assay kit by reading OD = 595 nm with FlexStation 3 and calculate protein concentration. Then prepare a protein sample with a reducing agent (1 mM DTT) and 4× LDS sample buffer, which is then heated at 70 °C for 10 min to denature proteins. Such samples can be stored at −80 °C for several months.
3. Following standard Western blotting technique, electrophorese equivalent amounts of protein (20–30 μg) on NOVEX NuPAGE 4–12 % Bis–Tris gels, and then electro-transfer to polyvinylidene difluoride (PVDF) membranes. Probe with primary antibodies (1:1,000 dilution) overnight at 4 °C followed by incubation with HRP-conjugated secondary antibody (1:5,000 dilution) for 45 min at room temperature.
4. Visualize protein levels by an enhanced chemiluminescent substrate, such as Western Lightning Plus-ECL (Fig. 3).

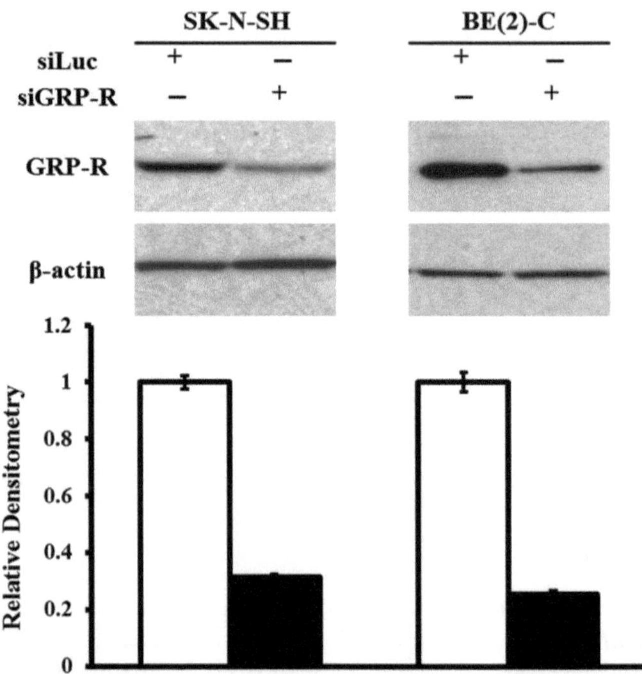

Fig. 3 GRP-R is silenced by SWNT–siRNA transfection. BE(2)-C and SK-N-SH cells were treated with SWNT–siRNA for 72 h. Protein expression was detected by Western blotting. GRP-R expression was significantly knocked down in both cell lines. β-actin levels indicate equal sample loading. Levels of protein expression were quantified by densitometry analysis with Image J (NIH)

3.8 SWNT–siRNA Delivery into Subcutaneous Xenograft Tumors In Vivo

All experiments were approved by the Institutional Animal Care and Use Committee in accordance with guidelines issued by the National Institutes of Health.

1. Maintain male athymic nude mice (4–6 weeks old) as described [16].
2. Establish BE(2)-C xenografts by injecting BE(2)-C cells (1×10^6) in 100 μl of HBSS into the bilateral flanks using a 26-gauge needle ($n = 3$–5 per group) as described [6, 16].
3. Perform SWNT–siRNA treatments from day 7 post tumor cell inoculation in mice by injecting 50 μl of the SWNT–siRNA PBS solution with concentrations at 20 mg/l for SWNTs and 1.7 μM for siRNA into each tumor.
4. Harvest tumors at the end point for analysis.

3.9 Immunohistochemistry

1. Perform immunohistochemical staining using DAKO EnVision + System-HRP kit.
2. Fix human neuroblastoma xenografts in formalin overnight and embed it in paraffin wax.
3. Mount tumor sections (5 μm) on glass slides.
4. Deparaffinize sections with xylene.

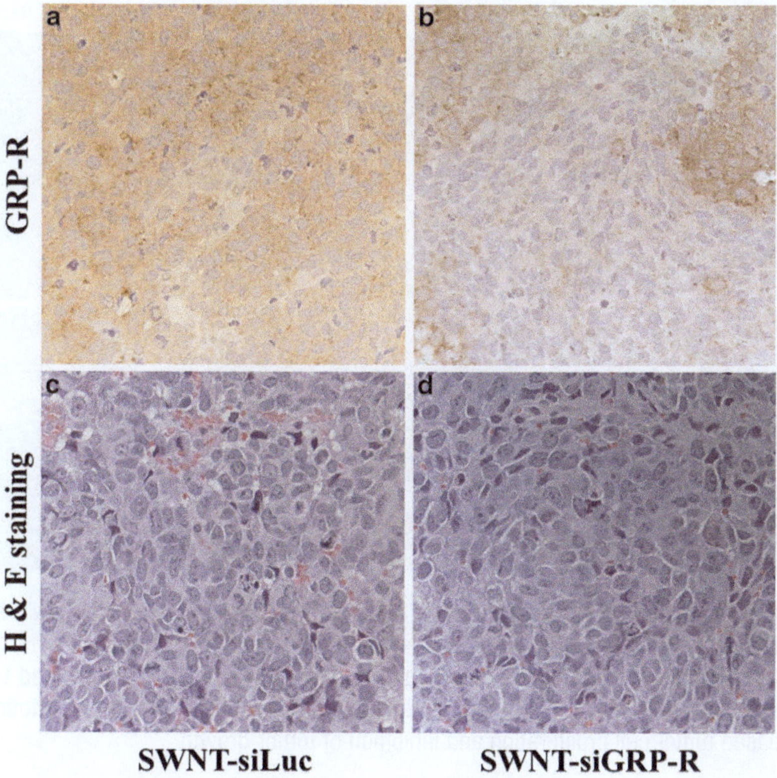

Fig. 4 SWNT–siGRP-R silencing in neuroblastoma in vivo. (**a**, **b**) Representative immunohistochemical staining of GRP-R in tumors treated with SWNT–siLuc or SWNT–siGRP-R. The expression of target GRP-R (*brown staining*) was significantly decreased in SWNT–siGRP-R-treated tumor sections. (**c**, **d**) Representative H&E-stained tumor sections from mice treated with SWNT–siLuc or SWNT–siGRP-R

5. Rehydrate sections with ethanol.
6. Perform antigen retrieval with 10 mM sodium citrate buffer.
7. Block sections with blocking solution for 20 min at room temperature.
8. Incubate slides with GRP-R primary antibodies overnight at 4 °C.
9. Wash slides with PBST buffer three times for 5 min each.
10. Incubate slides with secondary antibodies for 30 min at room temperature.
11. Develop sections with DAB reagent. The reaction was terminated by immersing slides in dH_2O, and sections were counterstained with hematoxylin.
12. Dehydrate slides with ethanol and xylene. Coverslips were mounted on slides.
13. Mount coverslips on the slides and leave them to dry.
14. Take the IHC images under microscope (Leica DMI6000 B) (Fig. 4).

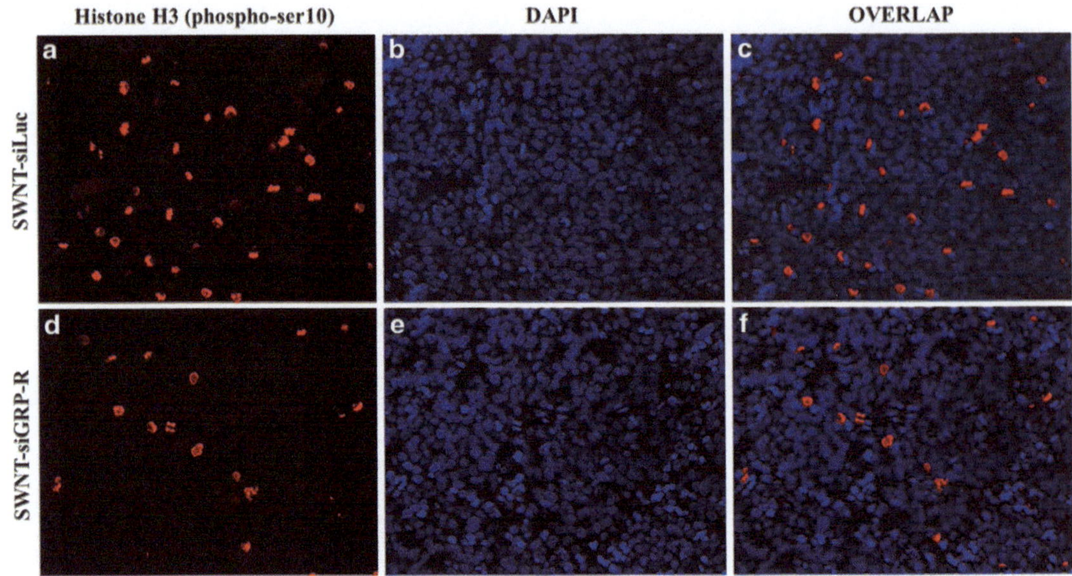

Fig. 5 Mitosis detected in paraffin-embedded tumor sections from SWNT–siLuc- and SWNT–siGRP-R-treated tumors, respectively. (**a**, **d**) Mitotic cells were stained with anti-human phospho-histone H3 (Ser10) primary antibody and Alexa Fluor 568 (*red*)-labeled secondary antibody. (**b**, **e**) Nuclei were stained with DAPI (*blue*). (**c**, **f**) Overlapped images. SWNT–siGRP-R treatments resulted in a reduced number of mitotic cells (**d** and **f**), leading to decreased tumor cell proliferation and inhibition of tumor growth

15. For mitosis detection, stain paraffin-embedded sections with rabbit polyclonal anti-human phospho-histone H3 (Ser10) antibody followed by Alexa Fluor 568 Dye (red)-labeled anti-rabbit secondary antibody staining.
16. Stain nucleus with DAPI (blue).
17. Capture images using a fluorescent microscope (Nikon Eclipse E600) (Fig. 5).

Acknowledgments

The authors would like to thank Nadja Colon and Roel L. Flores for critical review of the manuscript. This work was supported by R01 DK61470 (DHC) from the National Institutes of Health and ECCS-1055852 from National Science Foundation (YQX).

References

1. Kushner BH, Kramer K, LaQuaglia MP, Modak S, Yataghene K, Cheung NK (2004) Reduction from seven to five cycles of intensive induction chemotherapy in children with high-risk neuroblastoma. J Clin Oncol 22:4888–4892
2. Kim S, Hu W, Kelly DR, Hellmich MR, Evers BM, Chung DH (2002) Gastrin-releasing peptide is a growth factor for human neuroblastomas. Ann Surg 235:621–629, discussion 629–30

3. Moody TW, Carney DN, Cuttitta F, Quattrocchi K, Minna JD (1985) High affinity receptors for bombesin/GRP-like peptides on human small cell lung cancer. Life Sci 37:105–113
4. Markwalder R, Reubi JC (1999) Gastrin-releasing peptide receptors in the human prostate: relation to neoplastic transformation. Cancer Res 59:1152–1159
5. Qiao J, Kang J, Cree J, Evers BM, Chung DH (2005) Gastrin-releasing peptide-induced down-regulation of tumor suppressor protein PTEN (phosphatase and tensin homolog deleted on chromosome ten) in neuroblastomas. Ann Surg 241:684–691, discussion 691–692
6. Qiao J, Kang J, Ishola TA, Rychahou PG, Evers BM, Chung DH (2008) Gastrin-releasing peptide receptor silencing suppresses the tumorigenesis and metastatic potential of neuroblastoma. Proc Natl Acad Sci USA 105:12891–12896
7. Liu Z, Winters M, Holodniy M, Dai H (2007) siRNA delivery into human T cells and primary cells with carbon-nanotube transporters. Angew Chem Int Ed Engl 46:2023–2027
8. Kam NW, Liu Z, Dai H (2006) Carbon nanotubes as intracellular transporters for proteins and DNA: an investigation of the uptake mechanism and pathway. Angew Chem Int Ed Engl 45:577–581
9. Liu Z, Tabakman SM, Chen Z, and Dai H (2009) Preparation of carbon nanotube bioconjugates for biomedical applications. Nat Protoc 4:1372–1382
10. Schipper ML, Nakayama-Ratchford N, Davis CR, Kam NW, Chu P, Liu Z, Sun X, Dai H and Gambhir SS (2008) A pilot toxicology study of single-walled carbon nanotubes in a small sample of mice. Nat Nanotechnol 3: 216–221
11. Cherukuri P, Bachilo SM, Litovsky SH, Weisman RB (2004) Near-infrared fluorescence microscopy of single-walled carbon nanotubes in phagocytic cells. J Am Chem Soc 126:15638–15639
12. Cherukuri P, Gannon CJ, Leeuw TK, Schmidt HK, Smalley RE, Curley SA, Weisman RB (2006) Mammalian pharmacokinetics of carbon nanotubes using intrinsic near-infrared fluorescence. Proc Natl Acad Sci USA 103:18882–18886
13. Welsher K, Liu Z, Sherlock SP, Robinson JT, Chen Z, Daranciang D, Dai H (2009) A route to brightly fluorescent carbon nanotubes for near-infrared imaging in mice. Nat Nanotechnol 4:773–780
14. Tucker-Schwartz JM, Hong T, Colvin DC, Xu YQ, Skala MC (2012) Dual-modality photothermal optical coherence tomography and magnetic-resonance imaging of carbon nanotubes. Opt Lett 37:872–874
15. Hong T, Lazarenko RM, Colvin DC, Flores RL, Zhang Q, Xu Y-Q (2012) Effect of competitive surface functionalization on dual-modality fluorescence and magnetic resonance imaging of single-walled carbon nanotubes. J Phys Chem C 116:16319–16324
16. Chung DH, Kang JH, Ishola TA, Baregamian N, Mourot JM, Rychahou PG, Evers BM (2007) Bombesin induces angiogenesis and neuroblastoma growth. Cancer Lett 253: 273–281

Chapter 12

Amino Acid Mediated Linear Assembly of Au Nanomaterials

Manish Sethi and Marc R. Knecht

Abstract

Nanoparticles possess unique properties that are enhanced due to their small size and varied shapes. These properties can be directly manipulated by controlling the aggregation state, which can further be exploited for applications in bio/chemical sensing, plasmonics, and as supports for catalysts. While the advantages of controlled aggregates of nanomaterials are great, synthetic strategies to achieve such structures with precision over the final arrangement of the materials in three-dimensional space remain limited. We have shown that ligand exchange reactions on Au nanomaterials of various shapes using simple amino acids can induce the formation of linear aggregates of the materials. The assembly process is mediated by partial ligand exchange on the particle surface, followed by the surface segregation of the two ligands that produces an electric dipole across the nanomaterial from which alignment occurs in solution via dipole–dipole interactions. This linear-based assembly can be used to tune the optical properties of the materials and could represent new pathways to study the interactions between biological molecules and inorganic nanomaterials.

Key words Au nanoparticles, Au nanorods, Amino acids, Linear assembly

1 Introduction

Au nanomaterials possess distinct properties that include facile surface functionalization, high surface to volume ratios, and dynamic optical properties that are augmented based upon their solution aggregation state. These properties are based upon the quantum confinement effects that are achieved at the nanoscale [1–6]. The use of such materials for commercial and industrial applications is nearly limitless, where nanomaterials have made significant advances in catalysis, therapeutics, chemical and biological sensors, and energy storage [7–13]. While the activity of individual nanostructures remains important, the collection of such materials into controlled 3D arrangements represents new pathways to achieve structures with additional functionality. Thus, new methods to produce specifically designed structures through self-assembly processes to direct the fabrication of discrete arrangements is critically important. Asymmetric functionalization of the

Sandra J. Rosenthal and David W. Wright (eds.), *NanoBiotechnology Protocols*, Methods in Molecular Biology, vol. 1026,
DOI 10.1007/978-1-62703-468-5_12,

nanoparticle surfaces produces materials with a controlled ligand arrangement that can be used to fabricate ordered structures [14–18]; however, such strategies typically result in a low number of active materials. As such, methods that control the final aggregation state with a high degree of precision in a high throughput manner remain limited and must be developed to enhance assembly-based functional materials.

Recently, our research group has demonstrated that simple amino acids such as arginine and cysteine are able to linearly assemble both spherical Au nanoparticles and Au nanorods into branching, 1D nanochains. Using citrate-capped Au nanoparticles under selected reaction conditions, the secondary amino acid partially displaces the initial citrate molecules on the nanoparticle that leads to the formation of a patchy, charged surface due to segregation of the zwitterionic residues from the citrate molecules [17]. As a result, an electric dipole is generated across the nanoparticle that causes their linear assembly in solution [17]. This process can be directly controlled by varying the amino acid concentration, temperature, and solution dielectric, where selected aggregate sizes may be achieved [18].

Using similar biomolecule displacement-based methods, we have further demonstrated the linear assembly of anisotropically shaped Au nanorods. Such materials are particularly interesting due to their unique two-dimensional structure that results in the formation of two plasmon resonances: one associated with the longitudinal surface (LSP) and one from the transverse surface (TSP) [1]. The binding of cysteine at the Au nanorod tips assembles such structures into linear nanochains in a tip-to-tip arrangement. The amino acid preferentially binds at the tips due to the disruption of the surfactant bilayer that stabilizes these materials attributable to the high radius of curvature at this region [19]. As a result of cysteine attachment, the amine moieties of the α headgroup are exposed to solution, which can bind the tips of adjacent nanorods in solution in a pH dependent manner [16].

2 Materials

2.1 Au Nanoparticle Synthesis

1. Milli-Q water (18 MΩ cm; Millipore, Bedford, MA) is used for all experiments.
2. 19.69 mg of chloroauric acid (99.999 % $HAuCl_4·3H_2O$; Sigma-Aldrich, St. Louis, MO) is dissolved in 50.0 mL of deionized H_2O to prepare a 1.00 mM solution (*see* **Notes 1** and **2**).
3. Sodium citrate tribasic dihydrate (ACS reagent grade; Sigma-Aldrich; St. Louis, MO) is dissolved to a concentration of 38.8 mM in deionized water.

2.2 Au Nanorod Synthesis

1. Chloroauric acid (39.3 mg) is dissolved in 10.0 mL of deionized H_2O to prepare a 100 mM stock solution (*see* **Notes 1** and **2**).
2. $NaBH_4$ (≥98 %, 3.8 mg; EMD; Gibbstown, NJ) is dissolved in 10.0 mL of deionized H_2O to generate a 10.0 mM solution (*see* **Note 3**).
3. Ascorbic acid (176.13 mg; J.T. Baker; Phillipsburg, NJ) is dissolved in 10.0 mL of deionized H_2O.
4. A 10.0 mM silver nitrate (ACS Grade $AgNO_3$, VWR, Radnor, PA) solution is prepared by dissolving 17.0 mg of $AgNO_3$ in 10.0 mL of deionized H_2O. This solution container is wrapped in aluminum foil to avoid exposure to light, which can reduce Ag^+ to Ag^0.
5. Hexadecyltrimethylammonium bromide (CTAB) (3.6 g; ≥99.0 %, Sigma-Aldrich, St. Louis, MO) is dissolved in 100 mL of deionized H_2O (*see* **Notes 4** and **5**).

2.3 Linear Assembly of Au Nanomaterials

2.3.1 For Au Nanoparticles

L-arginine (871 mg, ≥98 %; Sigma-Aldrich; St. Louis, MO) is dissolved in 100 mL of deionized H_2O to prepare a 50.0 mM stock solution.

2.3.2 For Au Nanorods

L-cysteine (12.12 mg, ≥97 %; SAFC; Lenexa, KS) is dissolved in 1.00 mL of deionized, non-titrated H_2O to prepare a 100 mM stock solution.

3 Methods

3.1 Au Nanoparticle Synthesis via Citrate Reduction

1. Prior to synthesis, all glassware is thoroughly cleaned by soaking in aqua regia (HCl–HNO_3::3:1) for 20.0 min (*see* **Note 6**). The glassware is then extensively rinsed with deionized H_2O and dried in an oven at 80.0 °C.
2. Using a glass spatula, collect 19.69 mg of chloroauric acid in an Erlenmeyer flask and dissolve in 50.0 mL of deionized H_2O (*see* **Note 2**).
3. Place the flask on a stirring hot plate set at ~200 °C and attach a standard recirculating cold water condenser.
4. Heat the flask until the solution begins to reflux while vigorously stirring the solution.
5. Immediately after the solution begins to reflux, add 5.00 mL of the 38.8 mM sodium citrate tribasic solution at once. The solution color will change from bright yellow to colorless.
6. Reflux the mixture for 15.0 min as the solution color changes from colorless to dark red to bright red.

7. While keeping the flask connected to the condenser, bring the solution to room temperature slowly by removing the hot plate prior to use.
8. The synthesis of ~15.0 nm Au nanoparticles is confirmed by UV–Vis spectroscopy by monitoring the plasmon band at 520 nm.

3.2 Au Nanorod Synthesis

Au nanorods are prepared in a two step process using a seed mediated method in water [20].

3.2.1 Preparation of Au Seeds

1. Prior to synthesis, all glassware should be thoroughly cleaned by soaking in aqua regia ($HCl–HNO_3$::3:1) for 20.0 min (*see* **Note 6**). The glassware is then extensively rinsed with deionized H_2O and dried in an oven at 80.0 °C.
2. Using a glass spatula, add 39.3 mg of chloroauric acid to a glass vial and dissolve it in 10.0 mL of deionized H_2O (*see* **Note 2**).
3. To a 50.0 mL conical centrifuge tube, transfer a 250 μL aliquot of the chloroauric acid solution.
4. To this, add 7.50 mL of the 100 mM CTAB solution in water. The solution color will change from bright yellow to dull orange.
5. To the seed reaction, immediately inject 600.0 μL of the freshly prepared 10.0 mM ice cold solution of $NaBH_4$. The solution color will change to pale brown.
6. Leave the solution undisturbed for at least 2.00 h at 25.0 °C to ensure complete reduction (*see* **Note 7**).

3.2.2 Preparation of Au Nanorods

1. The approach described is for the generation of Au nanorods with an average diameter of 15 nm and an average length of 50 nm. To alter these dimensions, changes to the reaction seed concentration is used [20].
2. In a 50.0 mL conical centrifuge tube, add 2.00 mL of the 10.0 mM chloroauric acid stock solution.
3. To this, add 47.5 mL of the 100 mM CTAB stock solution. The solution color will change from bright yellow to dull orange.
4. Add 300 μL of the 10.0 mM freshly prepared silver nitrate solution carefully so as to avoid exposure of the silver nitrate to light. Mix by inversion twice.
5. Proceed immediately with the addition of 320 μL of freshly prepared 100 mM ascorbic acid. Mix by inversion three times, after which the solution will become colorless.
6. After the solution becomes colorless, add 210 μL of the freshly prepared seed solution (*see* **Note 8**).

7. Mix the solution gently by inverting the reaction ~10 times.
8. Leave the reaction mixture undisturbed for at least 3.00 h at room temperature. The solution color will change slowly over time from colorless to purple.
9. The nanorod synthesis is confirmed by UV–Vis where two distinct plasmon resonances are observed; a small band is noted at ~500 nm associated with the TSP, while a large peak at ~730 nm is observed for the LSP. The position of the LSP is controlled by the nanorod aspect ratio (length/width) such that as this ratio increases, the LSP shifts further to the red [20].

3.3 Arginine-Mediated Assembly of Citrate-Capped Au Nanoparticles

1. The arginine stock solution is added to 1.00 cm pathlength quartz cuvettes (3.50 mL volume) in different amounts as shown in Table 1. Such volumes are selected to achieve the desired arginine–Au nanoparticle ratio with a final reaction volume of 3.00 mL. (In Table 1, K = 1,000, e.g., 40K = 40,000). The arginine is then diluted with a sufficient volume of water such that only the Au nanoparticles must be added to initiate the reaction.
2. Using the Beer–Lambert Law, determine the concentration of the Au nanoparticle stock solution. The extinction coefficient of 15.0 nm Au nanoparticles is 3.6×10^8 cm^{-1} M^{-1} [21].
3. Add a sufficient volume of the Au nanoparticle stock to the arginine reaction so as to attain the final Au nanoparticle concentration of 2.00 nM. This volume is 600 μL (Table 1) when the nanoparticle stock concentration is 10.0 nM.
4. After nanoparticle addition, thoroughly mix the reaction solution using a pipette. After mixing, leave the reaction undisturbed during the UV–Vis study (~1.0–6.0 h).
5. Immediately after mixing the solution, begin monitoring changes in the optical properties of the materials via UV–Vis spectroscopy (Fig. 1). Due to the rate of optical changes, obtaining of a complete spectrum every 30 s is required using a photodiode array.
6. The color of the solution will change over time from red to blue to purple based upon the linear assembly of the materials as a function of the arginine concentration (Fig. 2). TEM analysis of the 40K sample (Fig. 3) shows the growth of linear assemblies of Au nanoparticles over 6.00 h.

3.4 Cysteine-Mediated Assembly of Au Nanorods

3.4.1 Preparation of Aqueous Solvents at Selected pH Values

1. Dilute 10.0 mL of concentrated HCl with 20.0 mL of deionized H_2O. This solution is used for titrating acidic solutions (*see* **Note 9**).
2. In a separate beaker, dissolve 8–10 pellets of NaOH in 20.0 mL of deionized H_2O. This solution is used for titrating neutral and basic solutions.

Table 1
Reagent volumes required to prepare the indicated arginine–Au nanoparticle ratio reaction solution

Arg:Au	[Arg] (μM)	Milli-Q H_2O (μL)	50 mM Arg (μL)	Au (μL)[a]
0K	0	2,400	0	600
10K	20	2,398.8	1.2	600
20K	40	2,397.6	2.4	600
40K	80	2,395.2	4.8	600
60K	120	2,392.8	7.2	600
80K	160	2,390.4	9.6	600
100K	200	2,388	12	600
200K	400	2,376	24	600
400K	800	2,352	48	600
1,000K	2,000	2,280	120	600
4,000K	8,000	1,920	480	600
8,000K	16,000	1,440	960	600

[a]Based upon a 10.0 nM Au nanoparticle stock solution

3. In a 100 mL beaker under vigorous stirring, titrate 50.0 mL of deionized H_2O using the solutions prepared above while closely monitoring the solution pH; add the HCl/NaOH solutions very slowly to avoid over-titration.

3.4.2 Preparation of Reaction Solution

1. Dissolve 12.1 mg of cysteine in 1.00 mL of deionized H_2O to prepare a 100 mM solution.
2. Centrifuge 1.00 mL aliquots of the crude Au nanorods in 1.50 mL microfuge tubes at 14,000 rpm for 10.0 min to pellet the materials. Carefully discard the supernatant without disturbing the pellet.
3. Redissolve the pellet in 1.00 mL of water titrated to the appropriate pH (*see* **Note 10**).
4. Transfer the solution to a 1.00 cm pathlength quartz cuvette with a total volume of 3.50 mL to monitor the reaction using UV–Vis.

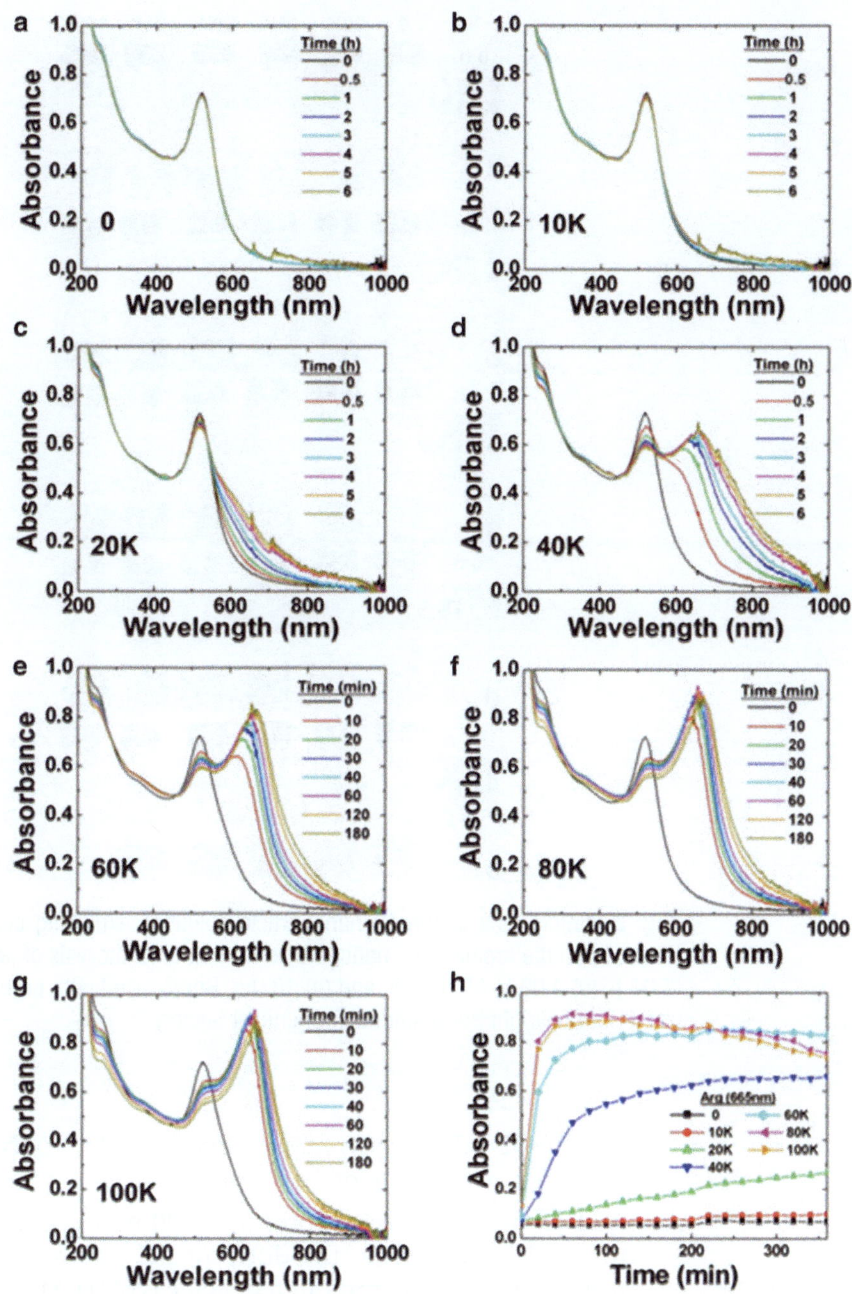

Fig. 1 UV–Vis analysis demonstrating the changes in the absorbance of Au nanoparticles based upon an arginine–Au nanoparticle ratio of (**a**) 0, (**b**) 10K, (**c**) 20K, (**d**) 40K, (**e**) 60K, (**f**) 80K, and (**g**) 100K. Part (**h**) displays the rate of production of the linear-assembly peak at 665 nm. Reproduced with permission from ref. [17]. Copyright 2009 American Chemical Society

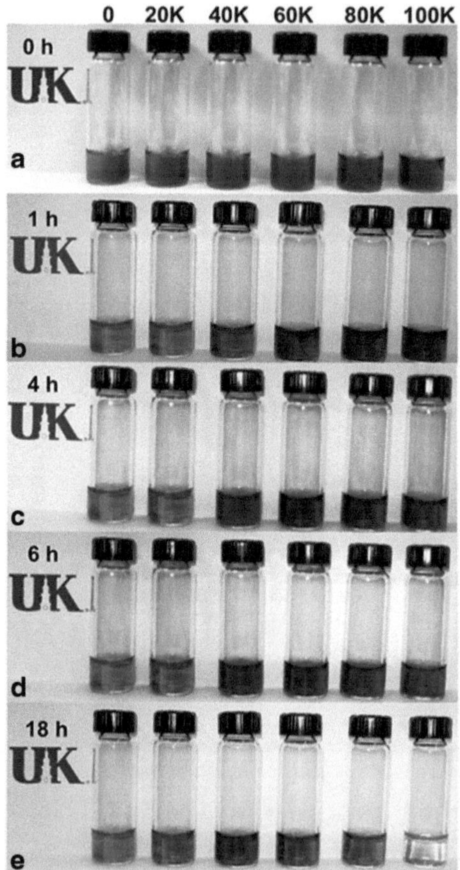

Fig. 2 Photographs of the Au nanoparticle solutions exhibiting color changes based upon the arginine–Au nanoparticle ratio at time intervals of (**a**) 0.00 h, (**b**) 1.00 h, (**c**) 4.00 h, (**d**) 6.00 h, and (**e**) 18.0 h. Reproduced with permission from ref. [17]. Copyright 2009 American Chemical Society

5. Add 1.982 mL of water at the same pH value to the solution to make a net volume of 2.982 mL.
6. Add 18 μL of the freshly prepared 100 mM cysteine solution to the Au nanorods at the different pH values such that the final reaction concentration of cysteine is 600 μM and the total volume is 3.00 mL.
7. Monitor the progress of the reaction every 3.00 min by UV–Vis spectroscopy for up to 4.00 h (Fig. 4).
8. Figure 5 presents the TEM images of cysteine-mediated, linearly assembled Au nanorods at pH 1.00, 2.00, and 3.00. No assembly is observed at higher pH values.

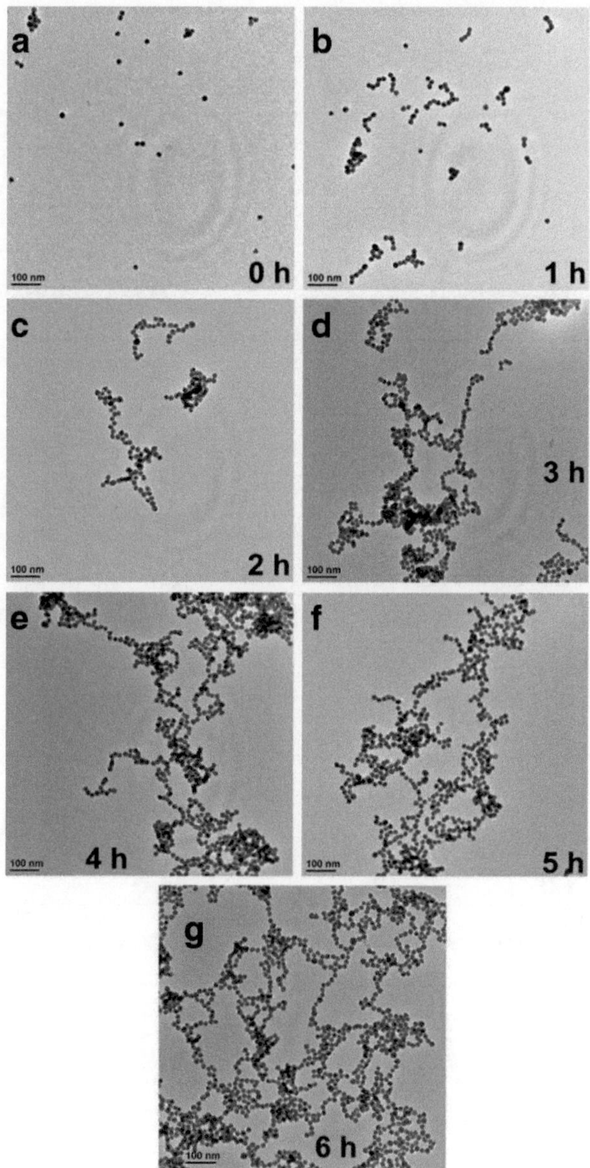

Fig. 3 TEM images of the 40K arginine–Au nanoparticle ratio sample at time points of (**a**) 0.00 h, (**b**) 1.00 h, (**c**) 2.00 h, (**d**) 3.00 h, (**e**) 4.00 h, (**f**) 5.00 h, and (**g**) 6.00 h. Reproduced with permission from ref. [17]. Copyright 2009 American Chemical Society

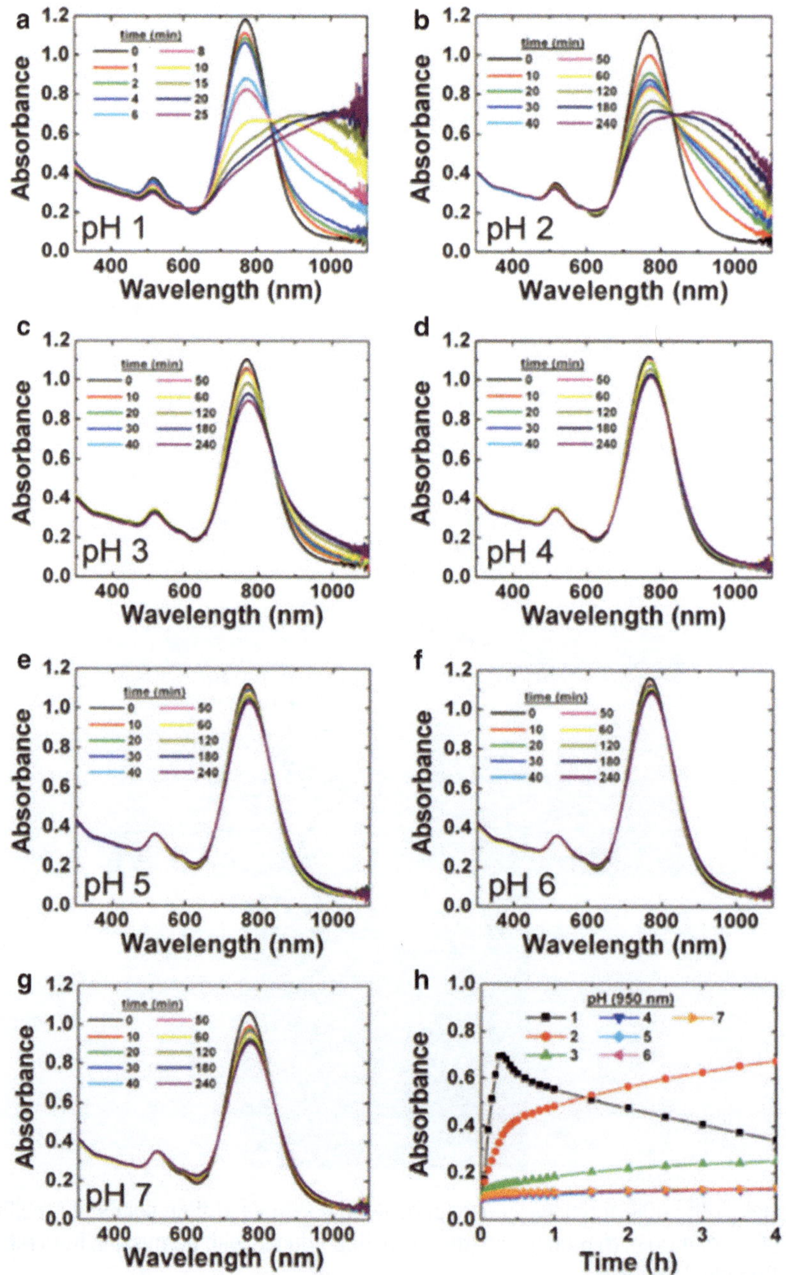

Fig. 4 UV–Vis analysis demonstrating differences in the Au nanorod assembly with the addition of cysteine at pH values of (**a**) 1.00, (**b**) 2.00, (**c**) 3.00, (**d**) 4.00, (**e**) 5.00, (**f**) 6.00, and (**g**) 7.00. Part (**h**) displays the rate of the LSP peak shifting at 950 nm for all pH values. Reproduced with permission from ref. [16]. Copyright 2009 American Chemical Society

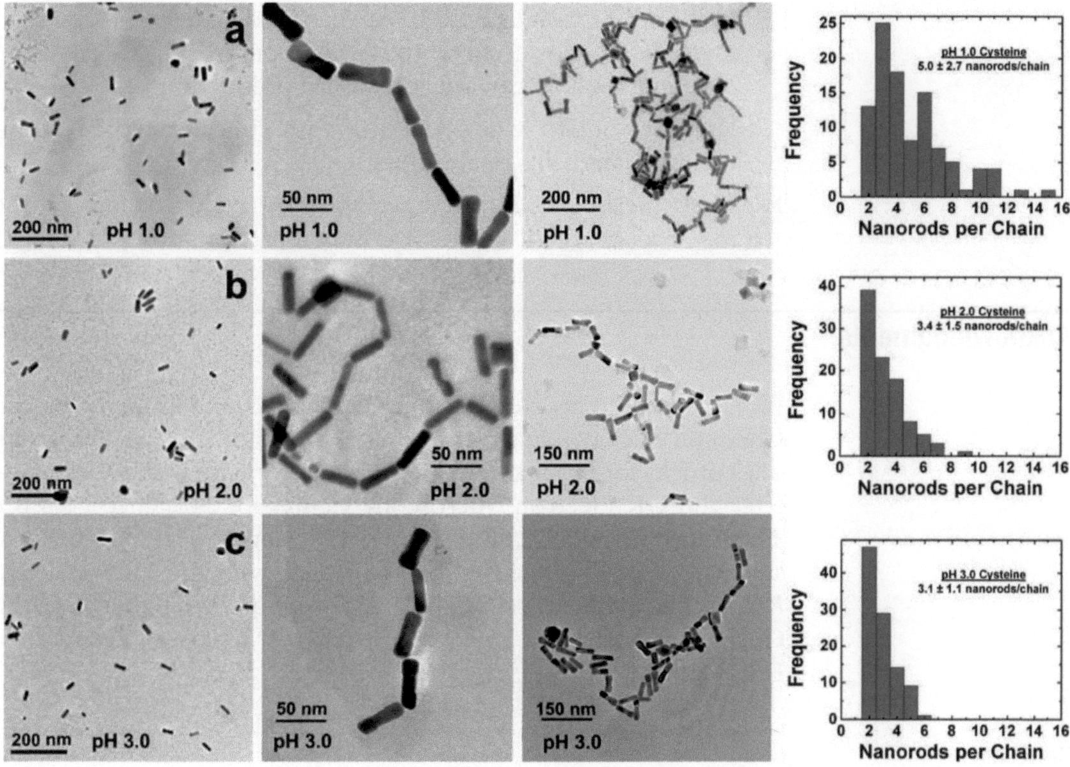

Fig. 5 TEM micrographs of the Au nanorods at pH (**a**) 1.00, (**b**) 2.00, and (**c**) 3.00 in the absence of cysteine (*left image*) and in the presence of cysteine (*middle* and *right images*). The histograms illustrate the number of nanorods per chain in the presence of cysteine for the corresponding pH sample. Reproduced with permission from ref. [16]. Copyright 2009 American Chemical Society

4 Notes

1. Chloroauric acid is highly hygroscopic and should be stored in a desiccator.
2. Metal spatulas should not be used with chloroauric acid as it reacts with the metallic surface. Ensure that glass-based spatulas are used.
3. $NaBH_4$ solutions must be prepared immediately before use.
4. Avoid frothing of CTAB solutions as much as possible.
5. Slightly heat the CTAB solution to fully dissolve the surfactant. Once dissolved, make sure the CTAB does not precipitate by preventing the solution from cooling below 25.0 °C.
6. Aqua regia is highly corrosive. Exercise caution with its handling and the associated chemical vapors.
7. Do not use the seed solution 6.00 h after preparation.

8. Using these conditions, if Au nanorods do not form after multiple reactions, check the CTAB; purchasing CTAB from different vendors can affect nanorod production [22, 23].
9. Exercise caution when working with concentrated HCl and the associated chemical vapors.
10. Wash the nanorods just once so as to avoid changes to the nanorod structure and their eventual precipitation [24].

Acknowledgments

We are grateful to the National Science Foundation, American Chemical Society—Petroleum Research Fund, and the University of Kentucky for financial support. The authors also wish to acknowledge the UK Electron Microscopy Center for assistance with the TEM analysis and Dr. T. Dziubla, UK Department of Chemical and Materials Engineering, for assistance with the DLS analysis. M.S. acknowledges student financial support from the Research Challenge Trust Fund and the University of Kentucky Presidential Fellowship.

References

1. Jain PK, Eustis S, El-Sayed MA (2006) Plasmon coupling in nanorod assemblies: optical absorption, discrete dipole approximation simulation, and exciton-coupling model. J Phys Chem B 110:18243
2. Jain PK, Huang X, El-Sayed IH, El-Sayed MA (2008) Noble metals on the nanoscale: optical and photothermal properties and some applications in imaging, sensing, biology, and medicine. Acc Chem Res 41:1578
3. Rosi NL, Giljohann DA, Thaxton CS, Lytton-Jean AKR, Han MS, Mirkin CA (2006) Oligonucleotide-modified gold nanoparticles for intracellular gene regulation. Science 312:1027
4. Daniel M-C, Astruc D (2004) Gold nanoparticles: assembly, supramolecular chemistry, quantum-size-related properties, and applications toward biology, catalysis, and nanotechnology. Chem Rev 104:293
5. Murphy CJ, Gole AM, Hunyadi SE, Stone JW, Sisco PN, Alkilany A, Kinard BE, Hankins P (2008) Chemical sensing and imaging with metallic nanorods. Chem Commun 544
6. Templeton AC, Wuelfing WP, Murray RW (2000) Monolayer-protected cluster molecules. Acc Chem Res 33:27
7. Rosi NL, Mirkin CA (2005) Nanostructures in biodiagnostics. Chem Rev 105:1547
8. Slocik JM, Naik RR (1988) Biologically programmed synthesis of bimetallic nanostructures. Adv Mater 2006:18
9. Pacardo DB, Sethi M, Jones SE, Naik RR, Knecht MR (2009) Biomimetic synthesis of Pd nanocatalysts for the stille coupling reaction. ACS Nano 3:1288
10. Slocik JM, Govorov AO, Naik RR (2008) Photoactivated biotemplated nanoparticles as an enzyme mimic. Angew Chem Int Ed 47:5335
11. Slocik JM, Zabinsky JS, Phillips DM, Naik RR (2008) Colorimetric response of peptide-functionalized gold nanoparticles to metal ions. Small 4:548
12. Knecht MR, Sethi M (2009) Bio-inspired colorimetric detection of Hg^{2+} and Pb^{2+} heavy metal ions using Au nanoparticles. Anal Bioanal Chem 394:33
13. Nam KT, Kim D-W, Yoo PJ, Chiang C-Y, Meethong N, Hammond PT, Chiang Y-M, Belcher AM (2006) Virus-enabled synthesis and assembly of nanowires for lithium ion battery electrodes. Science 312:885
14. Huo F, Lytton-Jean AKR, Mirkin CA (2006) Asymmetric functionalization of nanoparticles based on thermally addressable DNA interconnects. Adv Mater 18:2304

15. Xu X, Rosi NL, Wang Y, Huo F, Mirkin CA (2006) Asymmetric functionalization of gold nanoparticles with oligonucleotides. J Am Chem Soc 128:9286
16. Sethi M, Joung G, Knecht MR (2009) Linear assembly of Au nanorods using biomimetic ligands. Langmuir 25:1572
17. Sethi M, Knecht MR (2009) Experimental studies on the interactions between Au nanoparticles and amino acids: bio-based formation of branched linear chains. ACS Appl Mater Interfaces 1:1270
18. Sethi M, Knecht MR (2010) Understanding the mechanism of amino acid-based Au nanoparticle chain formulation. Langmuir 26:9860
19. Caswell KK, Wilson JN, Bunz UHF, Murphy CJ (2003) Preferential end-to-end assembly of gold nanorods by biotin-streptavidin connectors. J Am Chem Soc 125:13914
20. Sau TK, Murphy CJ (2004) Seeded high yield synthesis of short Au nanorods in aqueous solution. Langmuir 20:6414
21. Lee J-S, Stoeva SI, Mirkin CA (2006) DNA-induced size-selective separation of mixtures of gold nanoparticles. J Am Chem Soc 128:8899
22. Smith DK, Korgel BA (2008) The importance of the CTAB surfactant on the colloidal seed-mediated synthesis of gold nanorods. Langmuir 24:644
23. Smith DK, Miller NR, Korgel BA (2009) Iodide in CTAB prevents gold nanorod formation. Langmuir 25:9518
24. Sethi M, Joung G, Knecht MR (2009) Stability and electrostatic assembly of Au nanorods for use in biological assays. Langmuir 25:317

Chapter 13

Enzyme–Gold Nanoparticle Bioconjugates: Quantification of Particle Stoichiometry and Enzyme Specific Activity

Jacqueline D. Keighron and Christine D. Keating

Abstract

Enzyme–gold nanoparticle bioconjugates have a wide variety of uses ranging from nanoreactors to sensors and model systems. While easy to make, these bioconjugates are often not well characterized. This protocol outlines preparation of enzyme–nanoparticle bioconjugates and two complementary methods for quantifying enzyme:nanoparticle stoichiometry from which the specific activity of the adsorbed enzymes can be determined. Characterizations such as these can aid researchers in improving the design and application of future bioconjugate systems.

Key words Nanoparticle, Gold, Bioconjugate, Enzyme, Specific activity

1 Introduction

Enzyme–nanoparticle bioconjugates have a wide range of applications. They are commonly used as biocatalysts, where they can be used and reused as nanoreactors, as sensors where the optical or electrical properties of the nanoparticle enhance analyte detection [1], and as scaffolds for models of biological enzyme assemblies [2]. Direct adsorption of the enzyme to the surface of a nanoparticle is the simplest method for bioconjugate formation. When mixed with gold nanoparticles, protein molecules nonspecifically adhere to the nanoparticle surface through hydrophobic, van der Waals, and electrostatic interactions [3].

The orientation of adsorbed enzymes is most often random, meaning a portion of the enzymes adsorb with their catalytic center accessible to solution while others are less accessible because the catalytic center is blocked by the nanoparticle or neighboring enzymes. Inaccessibility of the active site, steric hindrance due to neighboring enzymes, and the loss of secondary and tertiary structure upon adsorption can lead to a significant loss in the catalytic ability of adsorbed species. The extent of activity loss is often unknown because the number of enzyme molecules bound per

Sandra J. Rosenthal and David W. Wright (eds.), *NanoBiotechnology Protocols*, Methods in Molecular Biology, vol. 1026,
DOI 10.1007/978-1-62703-468-5_13, © Springer Science+Business Media New York 2013

nanoparticle and hence the specific activity, or activity per quantity of enzyme, is not determined. Recently, efforts have been made to better characterize enzyme–nanoparticle bioconjugates in order to better understand the extent to which enzymes adsorb (i.e., a single layer or multiple layers around the particle) and how the enzyme activity changes upon adsorption [2, 4].

Two methods will be described for determining how many enzyme molecules adsorb per nanoparticle (also referred to as E:Au or particle stoichiometry). The first is a direct method in which the number of enzymes per nanoparticle is obtained after stringent washing steps, and the second is a more commonly used, indirect method, in which the stoichiometry is determined by quantifying the remaining unbound enzyme in solution. These methods are complementary, and when both are applied they provide a way to independently verify the measured stoichiometry. This insures an accurate determination in a system where enzyme–enzyme and enzyme–nanoparticle interactions, as well as interactions between the enzyme and any other surface or contaminant, can affect results.

While this protocol describes the characterization of fluorescently labeled citrate synthase–gold nanoparticle bioconjugates in depth, the protocol can be readily adapted for other enzymes of interest and other types of nanoparticles (i.e., silver or polystyrene).

2 Materials

1. Gold nanoparticles, for example, 30 nm diameter, 2.0×10^{11} particles/ml (Ted Pella Inc., Redding, CA).
2. Enzyme of interest, for example citrate synthase, 1 mg/ml stock solution.
3. Nonstick microcentrifuge tubes, 2.0 ml.
4. NaCl, 2 M.
5. Distilled deionized water.
6. Benchtop microcentrifuge.
7. Visible absorbance spectrophotometer.
8. Fluorescence spectrophotometer.
9. Enzyme substrate for citrate synthase, Oxaloacetic acid (Sigma-Aldrich, St. Louis, MO).
10. Enzyme substrate for citrate synthase, Acetyl-Coenzyme A (Sigma-Aldrich, St. Louis, MO).
11. Activity reporter for citrate synthase, dithionitrobenzoic acid (DTNB) (Sigma-Aldrich, St. Louis, MO).
12. Tris buffer (100 mM, pH 8.1).
13. Sodium bicarbonate (5 mM, pH 8.3).

14. Potassium cyanide (50 mM in 100 mM Tris buffer, pH 8) (*see* **Note 1**).
15. AlexaFluor protein label kit (Molecular Probes, Eugene, OR) (*see* **Note 2**).

3 Methods

Gold nanoparticles are available for purchase from a variety of sources including Sigma-Aldrich and Ted Pella Inc. in a large range of sizes. Nanoparticles can also be synthesized in house using well-established techniques [5]. Gold nanoparticles are produced by reduction of a gold salt solution and retain a coating of the reducing agent or an added stabilizing agent to prevent aggregation of the particles (*see* **Note 3**).

This protocol will outline the preparation and characterization of 30 nm enzyme–gold nanoparticle bioconjugates functionalized with citrate synthase. The procedure can be adapted for use with nanoparticles of any diameter (by accounting for the differences in nanoparticle surface area and concentration) and any enzyme of interest by performing the appropriate colorimetric activity assay for that enzyme (*see* **Note 4**).

3.1 Estimation of Monolayer Coverage

Initial estimations of number of enzymes necessary to form a monolayer on the surface of the nanoparticle can be made if the diameter of the enzyme and nanoparticle are known. These estimations can be used to predict how much enzyme will be necessary to coat the nanoparticle and if more than one layer of enzyme is present in the bioconjugate.

1. Determine the surface area of the nanoparticle using Eq. 1, where A_{NP} is the surface area of the nanoparticle and r_{NP} is the radius of the nanoparticle:

$$A_{NP} = 4\pi r_{NP}^2 \tag{1}$$

2. To roughly estimate the molecular footprint of each enzyme molecules on the particle surface, measure the average largest and smallest diameter of the enzyme of interest in Jmol (*see* **Note 5**) and determine the minimum and maximum area a single enzyme would occupy on the surface of the nanoparticle according to Eq. 2, where A_E is the area of the enzyme and r_E is the radius of the enzyme:

$$A_E = \pi r_E^2 \tag{2}$$

3. Alternatively, if a crystal structure for the enzyme of interest cannot be found, use Eq. 3 [6] to estimate the radius of the enzyme (r_E) by assuming the enzyme is spherical, where ν is

the specific volume of a protein (0.74 cm^3/g), m is the molecular weight of the enzyme, and N_A is Avogadro's number (*see* **Note 6**):

$$A_E = \pi\left(\frac{3\nu m}{4\pi N_A}\right)^{2/3} \quad (3)$$

4. The expected number of enzyme molecules in a single monolayer on the nanoparticles is then estimated by A_{NP}/A_E. Although this value is only a rough estimate, it is nonetheless useful to have some idea how many molecules one should expect to find in the bioconjugates.

3.2 Preparation of Enzyme Stock

The enzymes and labels used in these assays are chosen based on the desired application of the bioconjugates.

1. In order to quantify bioconjugate stoichiometry a colorimetric or fluorescent reporter is required. If employing a reporter such as a fluorophore covalently attached to the enzyme [7], follow the instructions provided with labeling kit to label the enzyme (*see* **Note 7**).
2. Solutions with an ionic strength greater than 5–10 mM can cause aggregation of gold nanoparticles, which will be visible to the eye as a color change from red to purple or blue or in the absorbance spectrum as a redshift in the optical absorbance spectrum. Use dialysis or filtration to remove high ionic strength buffers and replace them with a suitable low ionic strength alternative (*see* **Note 8**).

3.3 Flocculation Assays

A flocculation assay is a rapid colorimetric method for determining the ratio of enzyme necessary in solution to prevent salt-induced aggregation of gold nanoparticles [5]. While this method does not provide particle stoichiometry it does define the limits of bioconjugate stability.

1. Determine several possible enzyme to nanoparticle ratios necessary to coat the surface of the nanoparticle with enzyme. It is recommended to use a wide range of ratios, from tens to thousands of enzymes per nanoparticle (Table 1). A sample that does not contain enzyme should always be used as a control of the absorption spectra and appearance of gold nanoparticles. Multiple low ionic strength buffers may also be tested to determine the optimal adsorption pH (*see* **Note 9**).
2. In a microcentrifuge tube, add Au nanoparticles to appropriate amount of buffer and mix the samples by inverting the tube several times to ensure homogeneity as defined in Table 1. In order to accurately interpret the results, the concentration of nanoparticles in each sample should be identical. To account for the different volumes of enzyme stock added, buffer is added to bring each sample to the same final volume.

Table 1
Example flocculation ratios

CS:Au ratio	Vol. of Au (mol)	Vol. of CS (mol)	Vol. buffer (μl)
0:1	1,000 μl (3.32×10^{-13})	0 μl	500
50:1	1,000 μl (3.32×10^{-13})	1.41 μl (1.66×10^{-11})	498
100:1	1,000 μl (3.32×10^{-13})	2.82 μl (3.32×10^{-11})	497
500:1	1,000 μl (3.32×10^{-13})	14.11 μl (1.66×10^{-11})	485
1,000:1	1,000 μl (3.32×10^{-13})	28.22 μl (3.32×10^{-10})	472
5,000:1	1,000 μl (3.32×10^{-13})	141.1 μl (1.66×10^{-09})	359

3. Add appropriate amount of enzyme to each sample, mix well by inverting the microcentrifuge tube several times, and incubate samples for 1 h at 4 °C. The samples should be protected from light to prevent photobleaching of fluorescently labeled enzymes (*see* **Note 10**).
4. In order to accurately interpret the spectra of each sample they should be compared to the spectrum of bare, unmodified nanoparticles (i.e., 0 enzymes per 1 nanoparticle, referred to as the 0:1 ratio in this protocol). To the 0:1 ratio sample, add 150 μl deionized water. For each of the other samples, add 150 μl of 2 M NaCl. Mix each sample well by inverting the tube several times and incubate the samples for an additional 30 min at 4 °C protected from light.
5. With a UV–Vis spectrophotometer take the absorbance spectra of each sample between 400 and 800 nm. Using the 0:1 ratio sample as an exemplar for the spectrum of unaggregated nanoparticles, compare the spectrum of each sample. As seen in Fig. 1, the coverage conditions necessary to prevent aggregation are represented by the lowest enzyme:nanoparticle ratio that closely matches the spectrum of the 0:1 ratio sample. The addition ratio and adsorption buffer will be used in Subheading 3.3.

3.4 Bioconjugate Preparation and Purification

The addition ratio found in Subheading 3.3 determines the amount of enzyme that must be added to solution in order to have complete coverage of the nanoparticle. However, every enzyme added does not necessarily adsorb to the nanoparticle, leaving a portion of enzyme remaining in solution. To determine the actual enzyme:nanoparticle stoichiometry, all free enzyme must be first removed from the sample.

1. Using the addition ratio and adsorption buffer create at least five replicate samples by following **steps 2** and **3** in Subheading 3.3 (*see* **Note 11**).

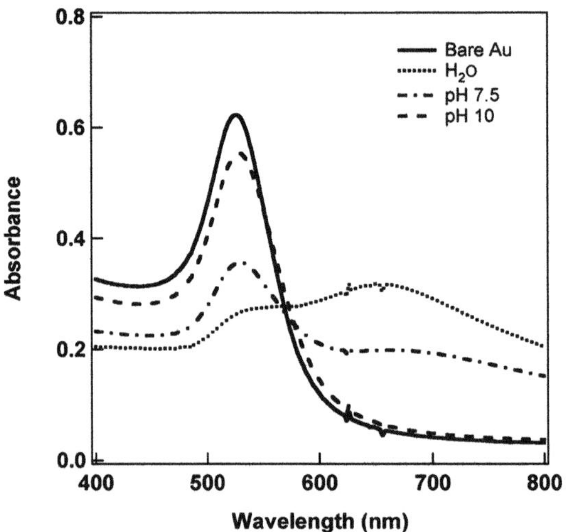

Fig. 1 Flocculation assay of citrate synthase with 30 nm Au in different pH buffers. The solid black trend represents the absorbance spectra of bare, unaggregated Au nanoparticles. Absorbance trends closely matching this (i.e., pH 10 solution) after incubation with NaCl are considered resistant to flocculation (aggregation), while trends which display a redshifted absorbance peak (i.e., H_2O) are considered unstable

2. Centrifuge each sample at 5,000 × *g* for 15 min at 4 °C (*see* Table 2 for alternative size nanoparticles), carefully remove the supernatant, and transfer to a new labeled centrifuge tube (*see* **Note 12**). Replace the volume removed with fresh adsorption buffer. Repeat this process three times (*see* **Note 13**).

3.5 Bioconjugate Activity

Activity assays are specific for each enzyme. Established assays are available with most purchased enzymes or in the primary literature. The general method for activity assays with citrate synthase bioconjugates as an example is presented here; this protocol will need to be modified for other enzymes of interest in accordance with literature or manufacturer protocols. Note that the activity of bioconjugates must be measured before quantifying stoichiometry (*see* **Note 14**).

1. Prepare a stock solution for each substrate. For citrate synthase, oxaloacetate (60 mM) and acetyl-CoA (12 mM) should be prepared in 100 mM Tris (pH 8.1). DTNB (18 mM) should be prepared in ethanol [2].
2. In a 3 ml quartz cuvette combine 2.725 ml of Tris buffer and 25 μl each of the oxaloacetate, acetyl-CoA, and DTNB solutions. Cap the cuvette and mix by inverting the cuvette several times [2]. Addition of bioconjugate (200 μL) is used to initiate the reaction.
3. Record the absorbance at 412 nm every 30 s for 5 min. Use the extinction coefficient for DTNB (13,600 M^{-1} cm^{-1}) [8] to

Table 2
Centrifuge conditions for bioconjugate purification

Nanoparticle diameter (nm)	RCF (1,000 × g)	Time (min)	Temperature (°C)
5	125	30	4
15–20	10	35	4
30–50	5	15	4

determine the concentration of product produced at each time point (for each mole of DTNB reacted, 1 mol of citrate is formed).

4. Repeat **steps 2** and **3** at least three times. Plot the results as seen in Fig. 2 and determine the activity, V (μmol/min), from the slope of the graph.

3.6 Indirect Quantification of Bioconjugate Stoichiometry

With proper control experiments the amount of enzyme bound to a single nanoparticle can be determined by comparing the known concentration of enzyme added to solution and the concentration of enzyme remaining in solution after adsorption occurs [3] (*see* **Note 15**).

1. Determine the amount of enzyme lost ($[E]_{lost}$) from solution by adsorption to the walls of the microcentrifuge tube by repeating the procedure laid out in Subheading 3.4 replacing the volume of nanoparticles added with buffer using Eq. 4, where $[E]_{total}$ is the initial amount of enzyme added to the nanoparticles during the adsorption step (mol), and $[E]_{sup}$ is the sum of amount of enzyme found in the supernatant samples (mol):

$$[E]_{lost} = [E]_{total} - [E]_{sup} \tag{4}$$

2. Determine the concentration of enzyme in each supernatant collected in Subheading 3.3 and **step 1** by measuring the fluorescence of each sample and comparing it to a calibration curve generated using dilutions of the stock enzyme solution.

3. Find the concentration of bioconjugates present by measuring the plasmon absorption spectra in a UV–Vis spectrophotometer using the extinction coefficient 3.585×10^9 M^{-1} cm^{-1} for 30 nm gold nanoparticles [9] (*see* **Note 16**).

4. Using Eq. 5 find the number of enzyme molecules adsorbed per nanoparticle (E:Au) for each replicate measurement, where $[E]_{total}$ is the initial amount of enzyme added to the nanoparticles during the adsorption step (mol), $[E]_{sup}$ is the sum of

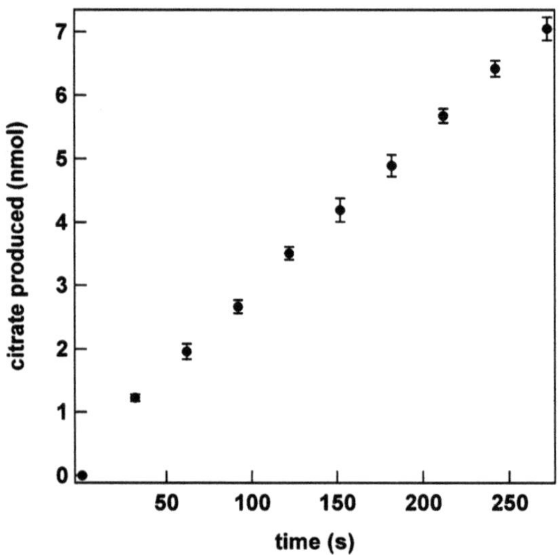

Fig. 2 Example activity assay for citrate synthase conjugates. The concentration of product formed is plotted against time, and the slope of this plot represents the activity of the adsorbed enzymes

amount of enzyme found in the supernatant samples (mol) found in **step 2**, $[E]_{\text{lost}}$ is the amount of enzyme lost to adsorption to the tube (mol) found in **step 1**, and [Bioconjugate] is the amount of bioconjugate (mol) determined in **step 3**:

$$E:\text{Au} = \frac{[E]_{\text{total}} - [E]_{\text{sup}} - [E]_{\text{lost}}}{[\text{Bioconjugate}]} \tag{5}$$

3.7 Direct Quantification of Bioconjugate Stoichiometry

Enzymes adsorbed to the surface of a gold nanoparticle can be released back into solution by dissolving the nanoparticle in a cyanide solution (*see* **Note 1**). This allows for fluorescence of the enzymes to be quantified (*see* **Note 17**).

1. To dissolve the gold nanoparticles, freeing the adsorbed enzyme molecules into solution: for each replicate measurement combine 300 μl of bioconjugate and 100 μl potassium cyanide buffer (*see* **Note 1**) in a microcentrifuge tube. Incubate for at least 3 h or until samples are optically clear (protect from light).
2. Prepare calibration standards. Due to the presence of several different species in the sample solutions, calibration standards for enzyme fluorescence must be incubated for the same time and under the same conditions (i.e., containing the same amount of bare gold nanoparticles and cyanide) as the bioconjugate samples.

3. Measure the fluorescence of each sample and use the calibration standards to determine the concentration of enzyme in each replicate sample. The bioconjugate stoichiometry (E:Au) can be calculated with Eq. 13.6, where $[E]_{\text{total}}$ is the amount of enzyme found (mol) and [Bioconjugate] is the amount of bioconjugate dissolved in each sample (mol):

$$E:\text{Au} = \frac{[E]_{\text{total}}}{[\text{Bioconjugate}]} \tag{6}$$

3.8 Determination of Specific Activity

Using the values collected in Subheading 3.5, and either Subheading 3.6 or 3.7, the specific activity of the adsorbed enzyme can be determined:

1. Calculate the average value and standard deviation for each measurement. If the stoichiometries found in Subheadings 3.6 and 3.7 are not the same within error repeat the measurements paying special attention to any unaccounted for loss of enzyme and any species that may interfere with fluorescence measurements. If the values found vary within error, then method with the lower % error should be used for specific activity determination.
2. Using Eq. 13.7 determine the specific activity (SA) of the bioconjugate in μmol/min, also called units (U), per mg of enzyme, where V is enzyme activity in μmol/min, [Bioconjugate] is the quantity of bioconjugate present in the activity (mol), E:Au is the number of enzymes per particle, and m is the molecular weight of the enzyme (g/mol):

$$\text{SA} = \frac{V}{\left([\text{Bioconjugate}] \times E:\text{Au} \times m \times 1{,}000\right)} \tag{7}$$

3. Bioconjugates are stable for approximately 1 week when stored at 4 °C protected from light. The rate of activity loss is dependent on the enzyme used and can be monitored through specific activity measurements (*see* **Note 18**).

4 Notes

1. The potassium cyanide solution used in this preparation must be adjusted to between pH 7 and pH 8. This should be done by slowly adding acid to the solution while monitoring the solution pH in a well-ventilated area. Acidic solutions will evolve cyanide gas, while basic solutions will interfere with fluorophores. For use with other diameter gold nanoparticles, a ratio of 16 cyanide molecules for each gold atom should be used [4].

2. Enzymes can be quantified several ways. Reporting reagents such as the Bradford reagent can be added to samples after the removal of nanoparticles. Alternatively, enzymes can be modified before adsorption with fluorescent or colorimetric reporters such as AlexaFluor dyes, fluorescein, and rhodamine. In this protocol enzymes are labeled with an AlexaFluor dye before adsorption to nanoparticles.
3. Some stabilization agents present on commercial gold nanoparticles can reduce the adsorption of enzymes to the particle surface. If this is a concern, washing the nanoparticles several times in 5 mM sodium bicarbonate (pH 8.3) or other suitable low ionic strength buffer can be used to remove excess stabilizing agent and improve the adsorption rate.
4. Activity assays may also employ other techniques such as fluorescence to quantify product formation.
5. The Protein Data Bank (www.rcsb.org/pdb/home/home.do) is a searchable database containing thousands of protein crystal structures. Programs to view protein structure and analyze structures such as Jmol, webmol, and Kingmol are available free of charge.
6. Most water-soluble proteins are globular, with relatively compact structures, making the spherical assumption reasonable for a rough estimate.
7. Covalent modifications to the surface of the enzyme can cause a loss of activity, particularly if amino acid residues near the active site are labeled [7]. Measure the activity of the enzyme of interest before and after labeling to determine the percent activity loss; if more than 50 % activity is lost consider a different labeling chemistry such as maleimide (which reacts with cysteine residues) instead of succinimide (which reacts with lysine).
8. Enzyme solutions can be sensitive to concentration. Stock solutions should be kept above 0.01 mg/ml enzyme to decrease loss of activity before adsorption.
9. When electrostatic effects are important in enzyme adsorption, adjusting the pH of the adsorption buffer to give the enzyme molecules a net positive charge can help drive adsorption to the negatively charged particles [10]. However, because protein molecules do not have uniformly distributed charge and other forces (hydrophobic interactions, van der Waals) are important in adsorption to gold nanoparticles, this is not a universal solution.
10. Over time fluorophores will photobleach and should not be kept in bright areas for long periods of time. For most enzymes storage at 4 °C or below helps prevent enzyme denaturation.

11. In some cases aggregation or excessive adsorption to the wall of the tube can occur at this step. If this occurs, try using a higher E:Au ratio.
12. The supernatants collected here will be used in the indirect quantification of nanoparticle stoichiometry.
13. It is common to lose a small portion of bioconjugate to adsorption to the tube and removal of the supernatant. If pellet is disrupted during centrifugation or no pellet is formed, recentrifuge samples and/or increase duration of centrifugation in 10 min increments.
14. Activity measurements are not reliable if bioconjugates have begun to aggregate. Perform an activity assay as soon as possible after preparing the bioconjugates so that any activity lost over time and/or due to aggregation and further enzyme denaturation can be observed later.
15. Indirect quantification is most reliable when the fraction of enzyme removed from solution is a significant portion of the total enzyme present. If only a tiny amount is removed this method will be less accurate.
16. To use this procedure with other types of nanoparticles determine what solvent will dissolve the particle and make sure to account for the effect of the solvent on the method chosen for enzyme quantification.
17. It is not feasible to quantify enzymes associated with a nanoparticle surface due to interferences from the nanoparticle absorbance spectra, fluorescence quenching by the metal [11], and scattering.
18. Over time, adsorption to the wall of the storage container can occur. Samples with significant loss of bioconjugates should be discarded even if aggregation in solution is not apparent.

Acknowledgment

This work was supported by the National Institutes of Health, grant R01GM078352.

References

1. Willner I, Bazaar B, Willner B (2007) Nanoparticle–enzyme hybrid systems for nanobiotechnology. FEBS J 274:302–309
2. Keighron JD, Keating CD (2010) Enzyme: nanoparticle bioconjugates with two sequential enzymes: stoichiometry and activity of malate dehydrogenase and citrate synthase on Au nanoparticles. Langmuir 26:18992–19000
3. Vertegal AA, Siegel RW, Dordick JS (2004) Silica nanoparticle size influences on the structure and enzymatic activity of adsorbed lysozyme. Langmuir 20:6800–6807
4. Cans A-S, Dean SL, Reyes FE, Keating CD (2007) Synthesis and characterization of enzyme–Au bioconjugates: HRP an fluorescein-labeled HP. NanoBiotechnology 3:12–22

5. Hayat MA (1989) Colloidal gold principles, methods, and applications, vol 1–3. Academic, San Diego, CA
6. Keating CD, Musick MD, Keefe MH, Natan MJ (1999) Kinetic and thermodynamics of Au colloid monolayer self-assembly: undergraduate experiments in surface and nanomaterials chemistry. J Chem Educ 76: 949–955
7. Hermanson GT (1996) Bioconjugate techniques. Academic, San Diego, CA
8. Santoro N, Brtva T, Vander Roest S, Siegel K, Waldrop GL (2006) A high-throughput screening assay for the carboxyltransferase subunit of acetyl-CoA carboxylase. Anal Chem 354:70–77
9. Ted Pella Inc. www.tedpella.com. Accessed May 2010
10. De Roe C, Courtoy PJ, Baudhuin P (1987) A model of protein–colloidal gold interactions. J Histochem Cytochem 35:1191–1198
11. Demers LM, Mirkin CA, Mucic RC, Reynolds RA III, Letsinger RL, Elghanian R, Viswanadham G (2000) A fluorescence-based method for determining the surface coverage and hybridization efficiency of thiol-capped oligonucleotides bound to gold thin films and nanoparticles. Anal Chem 72:5535–5541

Chapter 14

In Vivo Testing for Gold Nanoparticle Toxicity

Carrie A. Simpson, Brian J. Huffman, and David E. Cliffel

Abstract

A technique for measuring the toxicity of nanomaterials using a murine model is described. Blood samples are collected via submandibular bleeding while urine samples are collected on cellophane sheets. Both biosamples are then analyzed by inductively coupled plasma optical emission spectroscopy (ICP-OES) for nanotoxicity. Blood samples are further tested for immunological response using a standard Coulter counter. The major organs of interest for filtration are also digested and analyzed via ICP-OES, producing useful information regarding target specificity of the nanomaterial of interest. Collection of the biosamples and analysis afterward is detailed, and the operation of the technique is described and illustrated by analysis of the nanotoxicity of an injection of a modified tiopronin monolayer-protected cluster.

Key words Nanotoxicity, Murine model, Biodistribution, Tissue analysis, Organ digestion

1 Introduction

Nanotechnology represents a rapidly growing interface between the fields of chemistry and medicine. Currently nanomaterials are widely employed for in vivo imaging [1–3], targeting [4–6], and radiotherapy [7–9]. Significant advancements have been made with regard to biological mimicry with the eventual progression of vaccine generation through nanomaterials [10, 11]. Despite the synthetic progress, the toxicity of these nanomaterials has not been well studied. If these materials are to be used commercially, specific models for testing their toxicity must be established. One approach is to examine the effect of the nanomaterials using a controlled murine model.

A few classes of nanomaterials have been studied in vivo using murine models [12–16]; however, most of these models involve immediate euthanasia. Since the mice may be monitored for any short-term as well as long-term effects from the nanomaterial injection, the murine model is ideal for determining nanotoxicity. The simplest case would be to inject the mouse with the nanomaterial and observe any physical changes within the organism. However, given that the nanomaterial is absorbed into the bloodstream,

Sandra J. Rosenthal and David W. Wright (eds.), *NanoBiotechnology Protocols*, Methods in Molecular Biology, vol. 1026, DOI 10.1007/978-1-62703-468-5_14,

a multidimensional experiment may be performed to analyze for biodistribution, clearance rate, immune response, and more importantly targeting specificity of the desired nanomaterial.

Biodistribution and clearance rate may be calculated as a function of concentration of nanomaterial contained within the blood and urine at specific time points. If the concentration of nanomaterial within the urine is high immediately following injection, and dissipates, a relatively high clearance rate may be assumed. Most nanomaterials contain a metallic core; therefore, analytical analysis of the blood and urine for the specific metal should, in fact, quantify the concentration of the nanomaterial present within the sample. Likewise, analysis of the concentration of nanomaterial within the organs will yield targeting specificity data. Concentration of nanomaterial within the organs may be measured by reducing the organs in concentrated acid and analyzing for metal content.

In conjunction to biodistribution, the benefit of the murine model is the ability to deduce the immunogenicity provided by these nanomaterials. Many nanomaterials have been shown to produce specific binding and have potential as biomimics [17, 18]. However, it is unclear whether these nanomaterials would show the same effects in vivo. Unfortunately, not all components of the nanoparticle are susceptible to ELISA or immunogenic testing. In order to deduce whether these nanomaterials were producing any immunogenic response if ELISA or other immunogenicity tests are not applicable, a basic technique is available. Standard Coulter counters have been used for many years to determine red and white blood cell counts (RBC/WBC). Normal ranges have been established for almost every species and the equipment is extremely operator friendly. If a nanomaterial was inducing an immune response, an increase in WBC would be expected and is easily monitored.

The inductively coupled plasma optical emission spectroscopy (ICP-OES) analysis of gold concentration within blood and urine as a result of an injection of tiopronin monolayer-protected clusters (TMPCs) modified with polyethylene glycol (PEG) over the course of the entire murine model is shown in Fig. 1. As a result of the addition of PEG, the concentration of gold remained relatively consistent throughout the study within the bloodstream. The clearance rate was also noted as consistent throughout the study, implying the particle is escaping the process of opsonization, as has been noted in previous studies [19–21] as a hindrance to nanoparticle clearance rates. With the addition of PEG, an occurrence of an anti-PEG antibody at high concentrations has been shown [16, 20, 22]. Simple testing for the presence of this antibody has been noted in previous reports [23–25] and alternative methods for detection are currently being explored.

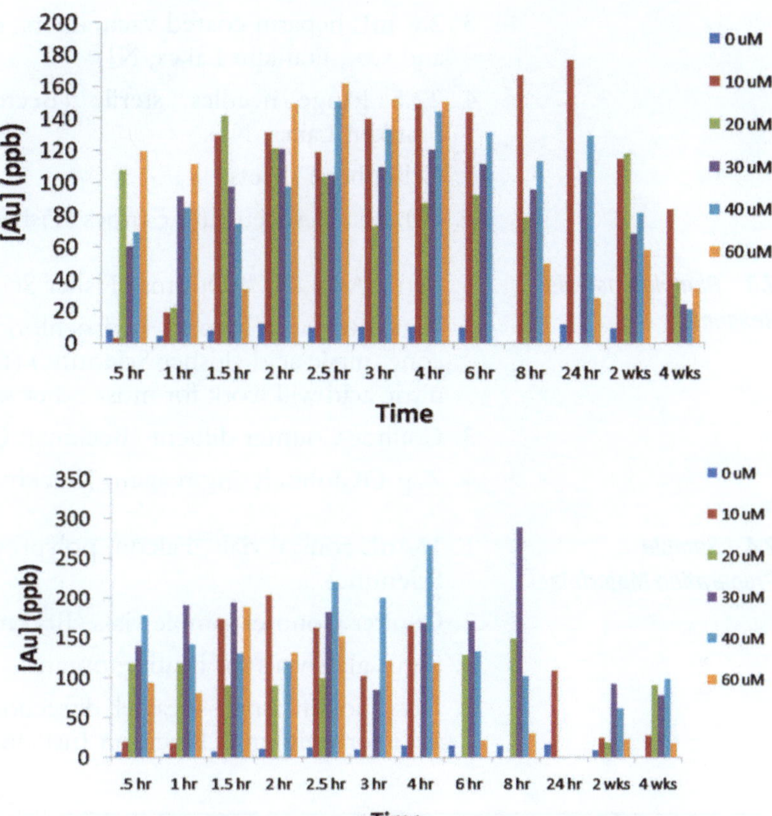

Fig. 1 ICP-OES data for gold concentration in blood (*top*) and urine (*bottom*) collected over the time course of proposed murine model after injection of tiopronin monolayer-protected clusters modified with polyethylene glycol. The addition of the PEG allows for prolonged residence time within the bloodstream, increasing the particles' candidacy for targeting specificity. As seen in the ICP-OES data, the particle remains at a constant concentration within the bloodstream even up to 2 weeks post injection until slowly declining at the 4-week sacrifice time point. The urine clearance rate also remains somewhat constant during the first 24 h. These findings have also been published elsewhere [26]

2 Materials

2.1 Equipment

1. Inductively Coupled Plasma-Optical Emission Spectrometer Optima DV 700 (Perkin Elmer) with a PC to run the instrument and accompanying software.
2. Coulter Counter and all corresponding materials necessary for measuring cell counts (Beckman Coulter).

2.2 Nanomaterial Injection/Specimen Collection Materials

1. 1 mL syringes, 27.5 gauge needles, sterile (Becton Dickinson and Co., Franklin Lakes, NJ).
2. Phosphate-buffered saline (PBS), sterile (Mediatech).

3. 2.0 mL heparin coated vacutainers, sterile (Becton Dickinson and Co., Franklin Lakes, NJ).
4. 22.5 gauge needles, sterile (Becton Dickinson and Co., Franklin Lakes, NJ).
5. Cellophane sheets.
6. 2.0 mL microcentrifuge tubes (Fisher Scientific).

2.3 Sample Analysis Reagents

1. Nitric Acid, 70 % Optima (Fisher Scientific).
2. Aqua Regia, 3:1 conc. hydrochloric acid (Fisher Scientific): conc. nitric acid (Fisher Scientific) [for gold analysis only; 2 % nitric acid will work for most other samples].
3. Coulter Counter diluent (Beckman Coulter).
4. Zap-OGlobin lysing reagent (Beckman Coulter).

2.4 Sample Preparation Materials

1. 15 mL conical vials, Falcon, polypropylene non-sterile (Fisher Scientific).
2. Coulter Counter sample vials (Beckman Coulter).
3. 3 mL glass vials for heating organs.
4. Dissection materials—scalpel, dissection board, scissors, tweezers, scale for weighing (Precision Instruments).

3 Methods

3.1 Animal Protocol

3.1.1 Species Specifications and Experimental Setup

1. A 6-week time course will be used for this model (Fig. 2) in which week 1 will include the baseline collection measurements, week 2 is the actual injection week, and weeks 4 and 6 serve as additional blood, urine, and tissue collection weeks, with eventual sacrifice at week 6 and full dissection. Organ harvesting of the spleen, heart, liver, and kidneys will be performed following sacrifice at week 6.
2. Unless breeding is desired, white, female, BALB/cAnNHsd 5–6-week mice should be chosen to limit aggressiveness. According to breeding guidelines, these mice should weigh between 15 and 20 g. Each mouse should be weighed prior to the beginning of the experiment (weight zero). This provides a tangible method of gauging nanotoxicity (*see* **Note 1**).
3. Allow no more than five mice per cage and label each mouse either by colored markings on their fur or by ear tagging.
4. Each cage should house mice injected with only one concentration of nanomaterial to avoid confusion. For example, cage 1 should only contain mice injected with 20 μM of particle.
5. After injection of the nanomaterial, the mice should be monitored frequently for the first 72 h. This includes frequent weighing, checking for muffled fur, and discoloration of the

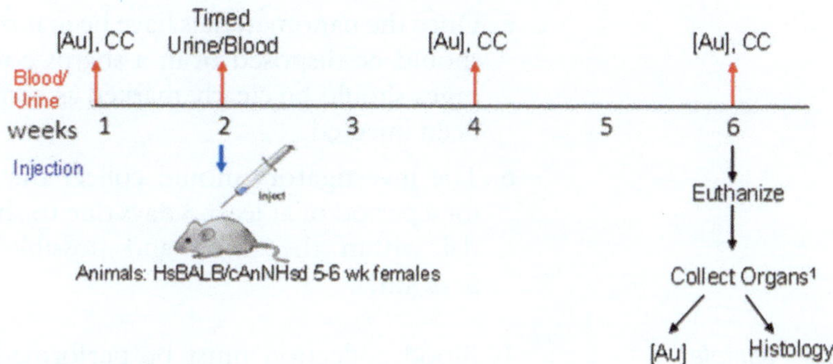

Fig. 2 Timeline for proposed murine model in which blood and urine are analyzed at 1-, 2-, 4-, and 6-week time points (shown here, analyzed for gold content) by ICP-OES. Immune response is monitored every 2 weeks by Coulter counter, with euthanasia 4 weeks post injection. Organs are then harvested for ICP-OES and histological analysis

front teeth, all of which are signs of distress. A loss of more than 15 % of the starting body weight at the time of injection is considered extreme distress; in this case, the animal should be immediately euthanized. Muffled fur and discoloration of the front teeth are also signs of extreme distress; in these cases, the animal is more than likely unable to eat and should be euthanized immediately. In cases where the experimenter is unsure, a veterinarian should be consulted immediately.

6. After the first 72 h, the animal may be considered to be stable, and the monitoring may become less frequent. Weights should be collected at least once a week to ensure the animal is still gaining weight and is not experiencing any adverse effects at the injection site.

3.1.2 Introduction of Nanomaterials

1. If possible, nanomaterials should be washed in ethanol and stored in a cool place prior to use to avoid bacterial contamination.
2. Nanomaterials must be suspended in sterile phosphate-buffered saline if possible or an equivalent sterile medium suitable for biological injection.
3. All syringes and needles must be sterile and laboratory personnel should wear appropriate PPE at all times when handling nanomaterials.
4. Nanomaterials may be injected either subcutaneously or intraperitoneally, although subcutaneously has been reported previously [26] to be the easiest and neither has been shown to give an advantage.

5. Once the nanomaterials have been introduced, the used needles should be disposed of in a sharps container, and the animals' cages should be clearly marked as containing animals that have been injected.
6. The investigators should collect the bedding from the cages for a period of at least 3 days due to shedding of the nanomaterial within the urine and possible hazard to non-trained personnel.

3.2 Sample Collection

3.2.1 Blood Collection

1. Blood collection must be performed at specific time points, dependent upon the number of mice available for the study. Typically, for ten mice/concentration, bleeds occur at 30 min, 1, 1.5, 2, 2.5, 3, 4, 6, 8, and 24 h during the week of injection. Blood samples are also taken 1 week prior and 2 and 4 weeks post injection.
2. Blood may be collected via the submandibular vein; this provides enough volume for all necessary analysis without sedation. Veterinary training on blood collection from this area may be necessary.
3. After puncturing the submandibular vein, one may immediately allow the blood to free-flow into the open heparin-coated tubes for collection until an adequate volume collected, making sure not to surpass the NIH bleeding guidelines.
4. The animal will most likely begin to clot immediately; however, clotting may be hastened by applying pressure to the wound with a small piece of sterile gauze.
5. Once the animal has clotted, it may be returned to the cage. Note that it may not be bled more than once a week (*see* **Note 2**).

3.2.2 Urine Collection

1. Urine collection should follow the same timeline as blood collection; however, urine may be collected as often as needed from the same mouse as many times as desired.
2. Secure a square of cellophane sheet approximately 6" × 6" and place it on a flat surface.
3. Gently place the mouse on the cellophane, making sure to secure the mouse by gently holding its tail.
4. In general, the mouse will void due to the texture of the cellophane; however, if the mouse is uncooperative, applying gentle pressure to the center of the back above the kidneys will usually cause the mouse to void immediately if possible (*see* **Note 3**).
5. Once the urine is present on the cellophane, immediately return the mouse to the cage to avoid contamination. Using a 3 mL plastic pipette, suction the urine off the cellophane and dispense into 2 mL microcentrifuge tubes for future analysis.

3.2.3 Organ Collection and Harvesting

1. If organ harvesting is desired for histological examination, it is best to dissect the animal immediately following euthanasia. The dissected organs may be placed in tissue cassettes and suspended in 10 % neutral buffered formalin solution for preservation until histological examination may be performed.
2. If organ harvesting is not a priority, the mice may be frozen at 4 °C until time permits for dissection. Allow adequate time for thawing before attempting to remove organs (*see* **Note 4**).
3. For examination of nanotoxicity, the major organs to be collected are those used in filtration and immunity: liver, spleen, kidney, and heart. These organs may be removed with a simple Y-incision and relatively few cuts.
4. When removing the kidneys, record the weights of both kidneys together, as they will be combined as one sample for ICP-OES. If one kidney is being used for histology, make note of this, and record only the weight of the ICP-OES sample kidney.
5. Removal of the liver can be challenging; recall that the liver has four lobes. The average mass of a whole murine liver was found to be approximately 1 g; if the results are less, the full sample may not have been excised.
6. The heart and spleen are relatively easy to remove; however, take great care in excising the spleen due to its small size (*see* **Note 5**).
7. Immediately following organ excision and weighing, all organs should be placed in individual 15 mL conical vials. Depending on the mass of the organ, an appropriate volume of 70 % Optima nitric acid should be added (*see* **Note 6**) to completely submerge the organ for digest. The organs should be untouched for at least 24 h.

3.3 Analysis of Serum and Tissue by Inductively Coupled Plasma Optical Emission Spectroscopy

3.3.1 Preparation of Blood and Urine Samples

1. Prepare 2 % nitric acid solution using 70 % Optima nitric acid in deionized water.
2. Prepare aqua regia solution (3:1 conc. hydrochloric acid:nitric acid) in glass beaker. Allow approximately 1.5 mL per sample to allow for loss during transfer and evaporation (*see* **Note 7**).
3. In a 15 mL non-sterile conical vial add 9 mL of 2 % nitric acid solution.
4. Draw 1 mL of the aqua regia solution; add to a small glass vial. To this, add 5 μL of whole blood/urine sample collected in heparin-coated vacutainers. Allow blood to dissolve in aqua regia.
5. Add aqua regia/blood solution to the 2 % nitric acid solution; gently mix.
6. Repeat this process for all blood and urine samples.

3.3.2 Digestion of Organ Samples

1. After the 24 h digestion in 70 % Optima nitric acid, the organs will be relatively dissolved. Transfer the contents of the conical vials to small glass vials, careful to label the vials before transfer (*see* **Note 8**).
2. Place small glass vials on low to medium heat (approximately 120 °C) and allow solutions to evaporate to dryness. Do not allow samples to char as this complicates transfer.
3. When samples have evaporated, remove from heat and allow cooling to RT. For gold core samples, add 1 mL aqua regia solution and swirl. If desired, light sonication may be applied to loosen debris.

3.3.3 Preparation of Organ Samples

1. Organ sample protocol is identical to blood and urine sample preparation.
2. In a 15 mL conical vial, combine 9 mL of 2 % nitric acid solution with 1 mL aqua regia/organ solution.
3. Repeat for all organ samples.

3.3.4 Parameters and Data Acquisition

Data acquisition will differ dependent upon the operating system of the ICP-OES of choice and element of interest. The following is a list of parameters for quantitative analysis of gold on a Perkin Elmer Optima DV 700 ICP-OES instrument.

1. Collect spectra in axial mode if possible as radial mode decreases sensitivity.
2. Set argon plasma flow to 15 L/min, nebulizer flow to 0.2 L/min, pump flow to 1.5 mL/min, and RF power at 1,300 W.
3. Choose a delay time of no less than 30 s; for our experiments, we chose 40 s.
4. For gold, choose the highest intensity wavelength for analysis, 267.595 nm.
5. For ICP-OES a three-point calibration is necessary; set these values to 1 ppm, 100 ppb, and 20 ppb respectively to fully cover the ranges observed for previous studies of gold particles. It will be necessary to adjust the calibration based upon the observed concentrations for the particular element(s) of interest after preliminary results are obtained.
6. Save method.

3.3.5 Correction Factors for Organ Analysis

Organ samples must be normalized according to weight for comparative study. Simply divide the elemental concentration attained from the ICP-OES (ppb/ppm) by the weight of each individual organ recorded previously. Report findings as ppb/g as shown in Fig. 3.

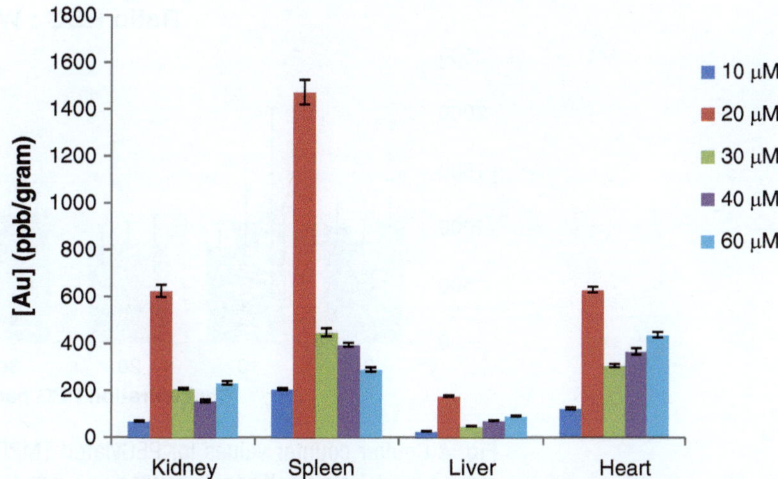

Fig. 3 ICP-OES organ data for PEGylated TMPCs normalized for comparison and given in ppb/g. Relatively high quantities of gold were shown within the kidney, spleen, and heart for all concentrations, and a correlating pattern may be noted for increasing concentration and increasing amounts of gold within the kidney, liver, and heart [26]

3.4 Immunological Assessment Utilizing a Standard Coulter Counter (Fig. 4)

3.4.1 Preparation of Red Blood Cell and White Blood Cell Samples

1. Into a standard Coulter counter vial, add 20 mL of isotonic diluent; label "WBC."
2. Into a second vial, add 19.8 mL of isotonic diluent; label "RBC."
3. To WBC vial, add 40 μL whole blood collected in heparin-coated vacutainer. Shake gently.
4. From WBC, transfer 200 μL to RBC.
5. Add 8–10 drops Zap-OGlobin lysing reagent to WBC. Shake gently. Solution will become clear and begin to turn orange.
6. Repeat these steps for all blood samples.

3.4.2 Data Acquisition and Analysis

1. Turn on Coulter counter; perform standard flush of system twice with isotonic diluent.
2. Set system to "counts"; gain = 256; $k_D = 60$; upper and lower size = 35 fL.
3. Record a blank (isotonic diluent with no sample); take this measurement three times and average the results. The result should be less than 100.
4. Proceed with samples; repeat the measurement three times per sample. Average all measurements for final result. This results in triplicate data analysis.

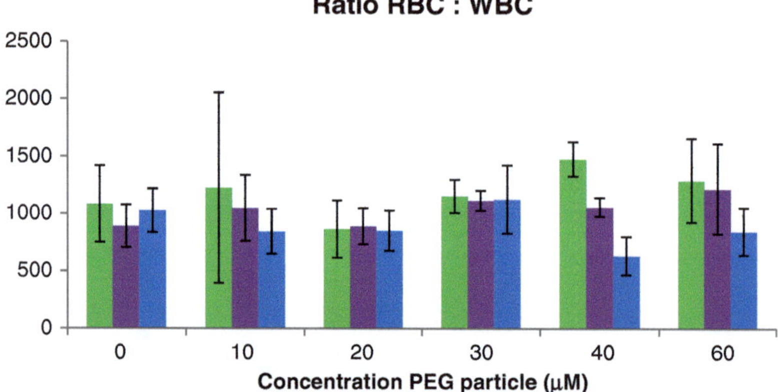

Fig. 4 Coulter counter values for PEGylated TMPC for weeks 1, 4, and 6 of the murine model. No significant immunogenic response is noted for 0–30 μM injections; all cell counts are within normal range for mice. There were significant WBC increases noted for the 40 and 60 μM injection species 2 and 4 weeks post injection, indicating the PEGylated TMPCs did, in fact, induce a form of immunity within the mouse at high concentrations. This work has also been published elsewhere [26]

5. Wash with diluent twice between samples to prevent contamination.
6. Results for RBC counts will be given in units of cells × 10^6/μL.
7. Results for WBC counts will be given in cells × 10^3/μL.
8. Report results as ratio of RBC:WBC (*see* **Note 9**).

4 Notes

1. Loss of more than 15 % of total body weight is indicative of extreme distress.
2. NIH guidelines suggest 7.5 % circulating blood volume maximum/week.
3. Use extreme caution when applying this method! A gentle touch is all that is needed to elicit voiding.
4. Usually 1 h in a well-ventilated hood space is adequate prior to dissection.
5. The spleen is a small tubular organ attached to the stomach. Use great care in removing to avoid cutting into pieces.
6. High purity nitric acid is essential for ICP-OES experiments; lower quality nitric acid will dissolve the samples but is not recommended for use in commercial instruments.

7. Use extreme caution when handling aqua regia solution! Extremely corrosive!
8. Label the vials before transfer if aqua regia is applied as the aqua regia will dissolve the labels.
9. Comparing individual counts is not feasible; there will be discrepancies in individual mice. However, the ratio of RBC:WBC will provide a fair, statistical data point for averaging purposes. Remember to account for the 3× magnitude difference in RBC and WBC counts when reporting the ratio.

References

1. Cherukuri P, Gannon CJ, Leeuw TK, Schmidt HK, Smalley RE, Curley SA, Weisman RB (2006) Mammalian pharmacokinetics of carbon nanotubes using intrinsic near-infrared fluorescence. Proc Natl Acad Sci USA 103:18882–18886
2. Stroh M, Zimmer JP, Duda DG, Levchenko TS, Cohen KS, Brown EB, Scadden DT, Torchilin VP, Bawendi MG, Fukumura D, Jain RK (2005) Quantum dots spectrally distinguish multiple species within the tumor milieu in vivo. Nat Med 11:678–682
3. Zimmer JP, Kim S-W, Ohnishi S, Tanaka E, Frangioni JV, Bawendi MG (2006) Size series of small indium arsenide-zinc selenide core-shell nanocrystals and their application to in vivo imaging. J Am Chem Soc 128: 2526–2527
4. Eck W, Craig G, Sigdel A, Ritter G, Old LJ, Tang L, Brennan MF, Allen PJ, Mason MD (2008) Pegylated gold nanoparticles conjugated to monoclonal F19 antibodies as targeted labeling agents for human pancreatic carcinoma tissue. ACS Nano 2:2263–2272
5. Everts M, Saini V, Leddon JL, Kok RJ, Stoff-Khalili M, Preuss MA, Millican CL, Perkins G, Brown JM, Bagaria H, Nikles DE, Johnson DT, Zharov VP, Curiel DT (2006) Covalently linked Au nanoparticles to a viral vector: potential for combined photothermal and gene cancer therapy. Nano Lett 6:587–591
6. Tkachenko AG, Xie H, Coleman D, Glomm W, Ryan J, Anderson MF, Franzen S, Feldheim DL (2003) Multifunctional gold nanoparticle-peptide complexes for nuclear targeting. J Am Chem Soc 125:4700–4701
7. Hainfeld JF, Slatkin DN, Smilowitz HM (2004) The use of gold nanoparticles to enhance radiotherapy in mice. Phys Med Biol 49:309–315
8. Hainfeld JF, Slatkin DN, Focella TM, Smilowitz HM (2006) Gold nanoparticles: a new X-ray contrast agent. Br J Radiol 79:248–253
9. Khlebtsov B, Zharov V, Melnikov A, Tuchin V, Khlebtsov N (2006) Optical amplification of photothermal therapy with gold nanoparticles and nanoclusters. Nanotechnology 17: 5167–5179
10. Taubert A, Napoli A, Meier W (2004) Self-assembly of reactive amphiphilic block copolymers as mimetics for biological membranes. Curr Opin Chem Biol 8:598–603
11. Ranney DF (2000) Biomimetic transport and rational drug delivery. Biochem Pharmacol 59:105–114
12. Choi HS, Liu W, Misra P, Tanaka E, Zimmer JP, Ipe BI, Bawendi MG, Fangioni JV (2007) Renal clearance of quantum dots. Nat Biotechnol 25:1165–1170
13. Moghimi SM, Porter CJH, Muir IS, Illum L, Davis SS (1991) Non-phagocytic uptake of intravenously injected microspheres in rat spleen: influence of particle size and hydrophilic coating. Biochem Biophys Res Commun 177:861–866
14. De Jong WH, Hagens WI, Krystek P, Burger MC, Sips AJAM, Geertsma RE (2008) Particle size-dependent organ distribution of gold nanoparticles after intravenous administration. Biomaterials 29:1912–1919
15. Sonavane G, Tomoda K, Sano A, Ohshima H, Terada H, Makino K (2008) In vitro permeation of gold nanoparticles through rat skin and rat intestine: effect of particle size. Colloids Surf B Biointerfaces 65:1–10
16. Armstrong JK, Meiselman HJ, Wenby RB, Fisher TC (2003) In vivo survival of poly(ethylene glycol)-coated red blood cells in the rabbit. Blood 102:94A
17. Gerdon AE, Wright DW, Cliffel DE (2005) Hemagglutinin linear epitope presentation on monolayer-protected clusters elicits strong

antibody binding. Biomacromolecules 6: 3419–3424

18. Gerdon AE, Wright DW, Cliffel DE (2006) Epitope mapping of the protective antigen of *B. anthracis* by using nanoclusters presenting conformational peptide epitopes. Angew Chem Int Ed Engl 45:594–598
19. Owens DE, Peppas NA (2006) Opsonization, biodistribution, and pharmacokinetics of polymeric nanoparticles. Int J Pharm 307:93–102
20. Alexis F, Pridgen E, Molnar LK, Farokhzad OC (2008) Factors affecting the clearance and biodistribution of polymeric nanoparticles. Mol Pharm 5:505–515
21. Moghimi SM, Hunter AC, Murray JC (2001) Long-circulating and target-specific nanoparticles: theory to practice. Pharmacol Rev 53: 283–318
22. Richter AW, Akerblom E (1984) Polyethylene glycol reactive antibodies in man: titer distribution in allergic patients treated with monomethoxy polyethylene glycol modified allergens or placebo, a Nd in healthy blood donors. Int Arch Allery Appl Immunol 74: 36–39
23. Armstrong JK, Hempel G, Koling S, Chan LS, Fisher TC, Meiselman HJ, Garratty G (2007) Antibody against poly(ethylene glycol) adversely affects PEG-asparaginase therapy in acute lymphoblastic leukemia patients. Cancer 110:103–111
24. Armstrong JK, Leger R, Wenby RB, Meiselman HJ, Garratty G, Fisher TC (2003) Occurrence of an antibody to poly(ethylene glycol) in normal donors. Blood 102:556A
25. Leger RM, Arndt P, Garratty G, Armstrong JK, Meiselman HJ, Fisher TC (2001) Normal donor sera can contain antibodies to polyethylene glycol (PEG). Transfusion 41:29S
26. Simpson CA, Huffman BJ, Gerdon AE, Cliffel DE (2010) Unexpected toxicity of monolayer protected gold clusters eliminated by PEG-thiol place-exchange reactions. Chem Res Toxicol 23:1608–1616

Chapter 15

Methods for Studying Toxicity of Silica-Based Nanomaterials to Living Cells

Yang Zhao, Yuhui Jin, Aaron Hanson, Min Wu, and Julia Xiaojun Zhao

Abstract

A number of silica-based nanomaterials have been developed in recent years. An important application of these nanomaterials is in the field of biological and biomedical applications. However, a major concern about the safety of the nanomaterials in vitro has been proposed. To address this problem, several approaches have been developed for a systematic investigation of the cytotoxicity and genotoxicity of silica-based nanoparticles. These methods are mainly based on the traditional toxicity study approaches but with some modifications. In this chapter, four important methods for studying of toxicity of silica-based nanomaterials are summarized. These methods can detect cell proliferation, cell viability, DNA damage, and the generation of reactive oxygen species (ROS). The protocols of each method are introduced in detail.

Key words Silica-based nanomaterials, Toxicity, Cell viability, Cell proliferation, DNA damage, Reactive oxygen species

1 Introduction

In past decades, the rapid development of nanotechnology has facilitated the synthesis and characterization of nanomaterials [1]. As a result, many new types of nanomaterials have been generated. These nanomaterials have demonstrated significant advantages in a wide variety of applications. One of the most promising applications is in the field of biological and biomedical area. For instance, the luminescent nanomaterials can be used as tags for identification and quantification of small amounts of biological targets [2–4]. However, due to the small size of nanomaterials, they can easily enter tissues or cells crossing through cytoplasma membrane. If the size is very small, the nanomaterials may enter cell nuclei. Consequently, the engineered nanomaterials interact with biological components in cells, and this interaction may cause unpredictable damage to living cells [5–7].

Sandra J. Rosenthal and David W. Wright (eds.), *NanoBiotechnology Protocols*, Methods in Molecular Biology, vol. 1026, DOI 10.1007/978-1-62703-468-5_15, © Springer Science+Business Media New York 2013

Many types of nanomaterials can be used to tag and monitor biological targets; however, silica nanoparticles exhibit unique traits that make it particularly useful as a matrix for those tags [8–10]. These unique properties include porosity, transparency, and low toxicity. The high porosity of amorphous silica nanoparticles provides the three-dimensional space required for the doping of functional components. Silica nanomaterials are apparently "transparent." They are unlikely to absorb light in the near-infrared, visible, and ultraviolet regions, which allows for the functional components inside silica matrix to keep their original optical properties. In addition, the well-established silica chemistry facilitates the surface modification of silica-based nanomaterials. Therefore, silica-based nanomaterials may have broader applications in the biological field compared to other nanomaterials in the future. Thus, the study of toxicity of silica-based nanomaterials is valuable for the advancement of modern biomedical and biological science.

According to literature work, several methods have been used to study the toxicity of silica-based nanomaterials. Usually these methods are adopted from traditional toxicity study approaches with modifications. Thus, these methods are not necessarily specific for evaluation of silica-based nanomaterials [11, 12]. Based on literature, the toxicity of silica-based nanomaterials can be examined in different aspects, such as cell proliferation, apoptosis, DNA integrity, and oxidative stress. In this chapter, four methods will be discussed, including MTT assay for testing cell proliferation, Vybrant assay for investigation of cell apoptosis, comet assay for evaluation of DNA integrity, and 2′,7′-dichlorodihydrofluorescein diacetate (H_2DCFDA) assay for detection of generation of reactive oxygen species (ROS).

MTT assay is the most common assay for testing of cell proliferation [13–16]. This assay is based on a colorimetric approach. The mitochondrial function of the cells is evaluated spectrophotometrically by measuring the extent of mitochondrial reduction of a tetrazolium salt, 3-(4,5-dimethylthiazol-2-yl)-2,5-diphenyltetrazolium bromide (MTT, yellow), to a formazan (purple). This reduction can only occur in living cells but not in dead cells, thus it can be used to determine cell proliferation.

In addition to the mitochondrial function, the permeability of the cell membrane can reveal whether or not the cells are healthy. The Vybrant assay can provide this information. In this assay, a specific dye molecule, YO-PRO-1, is employed to identify if the cells have undergone apoptosis. If apoptosis occurs, this dye molecule can permeate the cell membrane, and then the stain cells show the color of dye molecules. Meanwhile, necrosis can be identified by propidium iodide that is also used in the Vybrant assay. The propidium iodide can only label the cells that have undergone necrosis due to the disruption of nuclear membrane integrity. Therefore, the Vybrant assay is a sensitive approach for investigation of cell apoptosis and necrosis [17, 18].

The detection of integrity of DNA sequences in cells is an effective way to evaluate if the toxic effect occurs to living cells. Comet assay, which is also known as single-cell gel electrophoresis assay [17, 19–21], is commonly used for this study. When DNA is damaged, the DNA strands fragment into smaller pieces. After the DNA is denatured and run through gel electrophoresis, the small, broken off pieces of DNA will travel farther up. The gel under a fluorescence microscope will give the sample a comet shape. DNA that is not damaged will produce a single dot on gel electrophoresis as all the DNA is approximately the same size.

In order to measure cell stress, one should detect ROS that presents itself when a cell is under external stress. The chemical H_2DCFDA can be used as a probe for detection of ROS [22]. Inactive H_2DCFDA is non-fluorescent. When H_2DCFDA is internalized into cells, it is hydrolyzed to form H_2DCF by hydrolase. If ROS is present, H_2DCF is oxidized by the ROS to form DCF, which holds fluorescent properties. The fluorescence intensity is directly proportional to the amount of ROS in the cell. Although there are several methods in detecting ROS, only H_2DCFDA assay will be described in this chapter.

Overall, this chapter will only cover four methods that are focused on the in vitro study of the toxicity of silica-based nanomaterials. Other traditional methods for toxicity study can also be adopted for this evaluation. Changes in the protocols might be needed due to the difference between nanomaterials with bulk materials.

2 Materials and Instruments

2.1 Cell Culture and Nanoparticle Treatment

1. The cell culture medium was made of GIBCO RPMI 1640 medium (Invitrogen, NY) with 5 % newborn bovine serum (NBS, purchased from Hyclone, UT), 2 mM HEPES buffer (pH 7.2), and 100 U/mL penicillin and 100 μg/mL streptomycin (Invitrogen, NY). Water used in all experiments was ultra-purified (18.3 MΩ cm) using a Milli-Q Millipore system.
2. PBS tablet (MP Biomedicals, Solon, OH) was dissolved in ultra-purified water (1 tablet/100 mL).
3. Flat-bottom 96-well plate (Fisher Scientific Co., Pittsburgh, PA).

2.2 MTT Assay

1. MTT (3-(4,5-dimethylthiazol-2-yl)-2,5-diphenyl tetrazolium bromide) from Celltiter 96® nonradioactive cell proliferation assay kit was obtained from Promega (Madison, WI).
2. Stop solution: 10 % DMSO (Sigma-Aldrich, St. Louis, MO) and 10 % SDS (Invitrogen, NY) in PBS.
3. Multiskan spectrum high performance spectrophotometer plate reader (Thermo Electron Corporation).

2.3 Vybrant Assay

1. Glass bottomed culture dishes (Fisher Scientific Co., Pittsburgh, PA).
2. Vybrant apoptosis assay kit #4 (Invitrogen, NY).
3. PBS (*see* **item 2** in Subheading 2.1).
4. Hydrogen peroxide (H_2O_2) (Sigma-Aldrich, St. Louis, MO).
5. Carl Zeiss 510 META Confocal Fluorescence Microscope (Carl Zeiss MicroImaging, Inc, Thornwood, NY).

2.4 Comet Assay

1. CometSlide™ 2 well slide is from Trevigen, Inc. (Gaithersbur, MD).
2. Ca^{2+}, Mg^{2+} free PBS (MP Biomedicals, Solon, OH).
3. Lysis buffer: 0.3 M NaOH, 0.5 % natrium lauryl sarcosinate, 1.0 M NaCl, and 2.0 mM EDTA (rf Czene).
4. Normal Melting Agarose (NMAgarose), Low Melting Point agarose (LMPAgarose) (Sigma-Aldrich, St. Louis, MO).
5. TE buffer: 0.01 M Tris–HCl, 1 mM EDTA, pH 7.4.
6. DNA denaturing solution: 0.3 M NaOH containing 2 mM EDTA.
7. Alkaline electrophoresis solution: same as DNA unwinding solution.
8. 95 % Ethanol (reagent grade).
9. SYBR Green I staining solution: dilute the commercial SYBR® Green I solution (Trevigen Inc) by 10,000 times with TE buffer.

2.5 H_2DCFDA Assay

1. H_2DCFDA (Molecular Probes, Eugene, OR).
2. Dimethylsulfoxide (DMSO) (Sigma-Aldrich, St. Louis, MO).
3. PBS (Fisher Scientific, Pittsburgh, PA).
4. H_2O_2 (Sigma-Aldrich, St. Louis, MO).
5. Multiskan spectrum high performance spectrophotometer plate reader (Thermo Electron Corporation).

3 Methods

3.1 Cell Culture and Nanoparticle Treatment

1. Seed about 1×10^4 cells in each well of a 96-well plate in 200 μL of RPMI medium, and then incubate the cells overnight (37 °C, 5 % CO_2). Leave three wells empty for the blank control.
2. In regards to the phagocytosis property of the cells, add nanomaterials directly to the medium containing the cultured cells for 0.5–4 h. The time it takes for the nanomaterials to enter the cell is dependent on the nanomaterial's concentration and size.

3.2 MTT Assay [13–16]

1. After the nanoparticle treatment, add 20 μL of 5 mg/mL MTT solution to each well. Place on a shaker for 5 min to thoroughly mix the MTT with the medium.
2. Incubate the MTT solution with cells for 4 h at 37 °C to allow the MTT to be metabolized.
3. Add 100 μL of stop solution to each well. Place on a shaker for 2 h at room temperature to completely dissolve the formazan into the solvent.
4. Read absorbance at 570 nm using the plate reader, 670 nm can be used as reference wavelength.

3.3 Vybrant Assay [17, 18]

1. For a direct image, seed cells in the glass-bottomed culture dish.
2. Treat the cells with nanoparticles. Choose one dish to treat the cells with H_2O_2, as the positive control. And keep one dish without any treatment as the negative control.
3. Add 1 μL of YO-PRO-1 stock solution (from Vybrant apoptosis assay kit) and 1 μL of Propidium iodide (PI) stock solution to 1 mL of PBS and mix well.
4. Add 5–10 μL mixture solution to the glass bottom dish with 1 mL of medium in it. Mix and incubate the dish at 37 °C for 5 min.
5. Before placing under a confocal microscope, remove the medium and wash the cells with PBS once in order to remove extra dye molecules.
6. Add 1 mL of PBS, and take images of the cells under a confocal fluorescent microscope. For YO-PRO-1 dye, the excitation wavelength is 488 nm, and the emission wavelength is 520 nm. For PI dye, the excitation wavelength is 530 nm and the emission wavelength is 620 nm.

3.4 Comet Assay [17, 19–21]

Section 1: Prepare the slides

1. Prepare 1 % NMAgarose in PBS, and then microwave it until the agarose dissolved.
2. Clean the CometSlide with 70 % ethanol. While the NMAgarose is still hot, dip the slide into the melted NMAgarose. Wipe the underside of the slide to remove agarose, lay the slide in a tray on a flat surface, and let it dry. Store the slide at room temperature until needed.
3. Prepare 0.5 % LMPAgarose in PBS. Put the molten agarose in 37 °C water bath to cool and stabilize the temperature.

Section 2: Prepare the cell lysis

1. Prepare the lysis buffer and chill at 4 °C or on ice for at least 20 min before usage.

2. Remove the culture medium and gently wash the cells with PBS.
3. Scrape off the cells from the dish and resuspend the cells with 10 μL PBS.
4. Mix the resuspended cells at 1×10^5 cells/mL with LMPAgarose at a ratio at 1:10 (v/v), and immediately add 50 μL of cell mixture onto the Comet Slide. Use the side of the pipette tip to spread the mixture solution to make sure the slide is completely covered.
5. Place slides on a level plane at 4 °C (with the lights turned off) for 30 min. A 0.5 mm clear ring will appear at edge of the slide area.
6. Immerse the slides into precooled lysis buffer at 4 °C for 45 min.
7. Dry the extra buffer from slides and immerse them in a freshly prepared DNA denaturing solution (pH > 13), for 60 min at room temperature, and make sure the lights are off.
8. Place the slides into the electrophoresis tray, and add alkaline electrophoresis buffer until the buffer completely covers the slides.
9. Set the power supply, and then run the gel at 1 V/cm for 20 min.
10. Gently drain excess electrophoresis solution, and immerse twice in H_2O for 5 min each. Then immerse the slides in 70 % ethanol for 5 min.
11. Dry the samples at room temperature for 30 min.
12. Add 100 μL of SYBR green solution to each circle of the dried agarose. Wait for 5 min with the lights off.
13. Detect the DNA integrity with a florescence microscope at the wavelength of $\lambda_{ex}/\lambda_{em}$: 488 nm/520 nm.

3.5 H_2DCFDA Assay [11, 12, 22]

1. Seed about 1.0×10^4 cells in each well of 96-well plate. Incubate the cells at 37 °C and 5 % CO_2 overnight.
2. On the day of the experiment, freshly prepare the H_2DCFDA solution. In order to do this, dissolve the H_2DCFDA powder with a small amount of anhydrous DMSO. Then use PBS to dilute the dissolved H_2DCFDA.
3. Add the H_2DCFDA solution into the culture medium. Make the final concentration of H_2DCFDA to be 5 μM.
4. Incubate the cells with H_2DCFDA for 2 h at 37 °C.
5. Treat the cells with nanoparticles for 2–4 h in the incubator. Cells without any treatment are used as the negative control, and cells treated with H_2O_2 are used as the positive control.
6. Remove the medium and wash the cells with PBS.

7. Detect the fluorescence at excitation of 485 nm and emission of 520 nm using the Multiskan spectrum high-performance spectrophotometer plate reader.

The above protocols are variable when different size, shape, and type of silica-based nanomaterials are tested. Meanwhile, the cell line plays an important role since the ability of cellular uptaking nanoparticles varies with the type of cells. Therefore, the experimental conditions should be optimized by each individual researcher.

Acknowledgments

This work was supported by the National Science Foundation Grants CHE-0911472 and EPS-0814442.

References

1. Marquis BJ, Love SA, Braun KL, Haynes CL (2009) Analytical methods to assess nanoparticle toxicity. Analyst 134:425–439
2. Zhao X, Hilliard LR, Mechery JM, Wang Y, Jin S, Tan W (2004) A rapid bioassay for single bacterial cell quantization using bioconjugated nanoparticles. Proc Natl Acad Sci USA 101(42):15027–15032
3. Zhao X, Dytocio RT, Tan W (2003) Ultrasensitive DNA detection using highly fluorescent bioconjugated nanoparticles. J Am Chem Soc 125(38):11474–11475
4. Zhao X, Dytocio RT, Tan W (2003) Collection of trace amounts of DNA/mRNA molecules using genomagnetic nanocapturers. Anal Chem 75(14):3476–3483
5. Lin W, Huang YW, Zhou XD, Ma Y (2006) In vitro toxicity of silica nanoparticles in human lung cancer cells. Toxicol Appl Pharmacol 217:252–259
6. Liu X, Whitefield PD, Ma Y (2010) Quantification of F(2)-isoprostane isomers in cultured human lung epithelial cells after silica oxide and metal oxide nanoparticle treatment by liquid chromatography/tandem mass spectrometry. Talanta 81: 1599–1606
7. Donaldson K, Stone V, Tran CL, Kreyling W, Borm PJA (2004) Nanotoxicology. Occup Environ Med 61:727–728
8. Jin Y, Li A, John CL, Hazelton SG, Liang S, Selid PD, Pierce DT, Zhao JX (2010) Amorphous silica nanohybrids: synthesis, properties and applications. Coord Chem Rev 253:2998–3014
9. Liang S, Hartvickson S, Kozliak E, Zhao JX (2009) Effect of amorphous silica nanomatrix on kinetics of metallation of encapsulated porphyrin molecules. J Phys Chem C 113: 19046–19054
10. Xu S, Hartvickson S, Zhao JX (2008) Engineering of SiO_2-Au-SiO_2 sandwich nanoaggregates using a building block: single, double and triple cores for enhancement of near infrared fluorescence. Langmuir 24(14): 7492–7499
11. Zhang Q, Matsuzaki I, Chatterjee S, Fisher AB (2005) Activation of endothelial NADPH oxidase during normoxic lung ischemia is KATP channel dependent. Am J Physiol Lung Cell Mol Physiol 289:954–961
12. Wan R, Mo Y, Zhang X, Chien S, Tollerud DJ, Zhang Q (2008) Matrix metalloproteinase-2 and -9 are induced differently by metal nanoparticles in human monocytes: the role of oxidative stress and protein tyrosine kinase activation. Toxicol Appl Pharmacol 233:276–285
13. Mossman T (1983) Rapid colorimetric assay for cellular growth and survival: application to proliferation and cytotoxicity assays. J Immunol Methods 65:55–63
14. Carmichael J, Degraff WG, Gazdar AF, Minna JD, Mitchell JB (1987) Evaluation of a tetrazolium-based semi-automated colorimetric assay: assessment of chemo sensitivity testing. Cancer Res 47:936–942
15. Sayes CM, Wahi R, Kurian PA, Liu Y, West JL, Ausman KD, Warheit DB, Colvin WL (2006) Correlating nanoscale titania structure with

toxicity: a cytotoxicity and inflammatory response study with human dermal fibroblasts and human lung epithelial cells. Toxicol Sci 92:174–185

16. Hussain SM, Javorina AK, Schrand AM, Duhart HM, Ali SF, Schlager JJ (2006) The interaction of manganese nanoparticles with PC-12 cells induces dopamine depletion. Toxicol Sci 92:456–463
17. Jin Y, Kannan S, Wu M, Zhao JX (2007) Toxicity of luminescent silica nanoparticles to living cells. Chem Res Toxicol 20:1126–1133
18. Fröhlich E, Samberger C, Kueznik T, Absenger M, Roblegg E, Zimmer A, Pieber TR (2009) Cytotoxicity of nanoparticles independent from oxidative stress. J Toxicol Sci 34:363–375
19. Singh NP, McCoy MT, Tice RR, Schneider EL (1988) A simple technique for quantitation of low levels of DNA damage in individual cells. Exp Cell Res 175:184–191
20. Olive PL, Wlodek D, Durand RE, Banáth JP (1992) Factors influencing DNA migration from individual cells subjected to gel electrophoresis. Exp Cell Res 198:159–267
21. Czene S, Testa E, Nygren J, Belyaev I, Harms-Ringdahl M (2002) DNA fragmentation and morphological changes in apoptotic human lymphocytes. Biochem Biophys Res Commun 294:872–878
22. Al-Mehdi AB, Zhao G, Dodia C, Tozawa K, Costa K, Muzykanotv V, Ross C, Blecha F, Dinauer M, Fisher AB (1998) Endothelial NADPH oxidase as the source of oxidants in lungs exposed to ischemia or high K^+. Circ Res 83:730–737

INDEX

Sandra J. Rosenthal and David W. Wright (eds.), *NanoBiotechnology Protocols*, Methods in Molecular Biology, vol. 1026, DOI 10.1007/978-1-62703-468-5, © Springer Science+Business Media New York 2013

O

P

Q

R

S

T

V

MIX
Papier aus verantwortungsvollen Quellen
Paper from responsible sources
FSC® C105338

If you have any concerns about our products,
you can contact us on
ProductSafety@springernature.com

In case Publisher is established outside the EU,
the EU authorized representative is:
Springer Nature Customer Service Center GmbH
Europaplatz 3, 69115 Heidelberg, Germany

Printed by Libri Plureos GmbH
in Hamburg, Germany